ROBOTICS APPLICATIONS FOR INDUSTRY

A Practical Guide

by

L.L. Toepperwein, M.T. Blackmon, W.T. Park

General Dynamics Corporation
Fort Worth, Texas

W.R. Tanner

Productivity Systems Inc.
Farmington, Michigan

and

W.F. Adolfson

Advanced Technology, Inc.
McLean, Virginia

NOYES DATA CORPORATION
Park Ridge, New Jersey, U.S.A.
1983

Library of Congress Catalog Card Number: 83-13109
ISBN: 0-8155-0962-6
Printed in the United States

Published in the United States of America by
Noyes Data Corporation
Mill Road, Park Ridge, New Jersey 07656

10 9 8 7 6 5 4 3 2 1

Library of Congress Cataloging in Publication Data

Main entry under title:

Robotics applications for industry.

Bibliography: p.
Includes index.
1. Robots, Industrial. I. Toepperwein, L. L.
TS191.8.R58 1983 629.8'92 83-13109
ISBN 0-8155-0962-6

Foreword

This book, based on studies by General Dynamics Corporation, Productivity Systems Inc. and Advanced Technology Inc., provides an introduction to the basic concepts and techniques pertaining to the use of robotics technology for manufacturing and industrial applications. A special section of the book is devoted to robotics in motor vehicle manufacture.

Industrial robots are multifunctional, programmable devices which manipulate and transport manufacturing components in order to perform manufacturing tasks, tasks too physically demanding, menial, or repetitive for a man to do efficiently. The robots generally consist of an arm, to which an end effector (gripper, spot welder, drill) is affixed; a power source; and a control unit providing logical direction for the entire unit.

In an era when cost-effective technology and increased productivity are of prime concern, these studies should be of value in many decision-making processes. There is little question that computer aided manufacturing (CAM) is likely to become the technology of choice in coming years, and that one of its first major U.S. applications has been motor vehicle manufacture.

The book contains two parts. Part I details the basic concepts of robotics technology—robot configuration, sensors, tooling, work station integration, current application information, and future trends. It provides an approach to the implementation of the technology, a means of determining implementation costs, and a review and documentation of available literature. Part II covers robotics use in motor vehicle manufacture. Patterns of usage and current robot populations are given for the worldwide auto industry. New technology development requirements are identified and current research efforts described.

The information in the book is from the following documents:

ICAM Robotics Applications Guide (RAG), prepared by L.L. Toepperwein, M.J. Blackmon, and W.T. Park of General Dynamics Corporation for Materials Laboratory, Air Force Wright Aeronautical Laboratories, Wright-Patterson Air Force Base, Ohio, April 1980.

Robotics Use in Motor Vehicle Manufacture, prepared by W.R. Tanner of Productivity Systems Inc. and W.F. Adolfson of Advanced Technology, Inc. for the U.S. Department of Transportation, February 1982.

The table of contents is organized in such a way as to serve as a subject index and provides easy access to the information contained in the book. An extensive and thorough glossary of robotics terminology is also included.

Notice

This technical report has been reviewed and is approved for publication. When Government drawings, specifications, or other data are used for any purpose other than in connection with a definitely related Government procurement operation, the United States Government thereby incurs no responsibility nor any obligation whatsoever; and the fact that the government may have formulated, furnished, or in any way supplied the said drawings, specifications, or other data, is not to be regarded by implication or otherwise as in any manner licensing the holder or any other person or corporation, or conveying any rights or permission to manufacture, use, or sell any patented invention that may in any way be related thereto.

Publication does not signify that the contents necessarily reflect the views and policies of the contracting agencies or the publisher, nor does mention of trade names or commercial products constitute endorsement or recommendation for use.

Contents and Subject Index

PART II: ROBOTICS IN MOTOR VEHICLE MANUFACTURE

Part I

Robotics Application Technology—Basic Concepts

The information in Part I is from *ICAM Robotics Application Guide (RAG)*, prepared by L.L. Toepperwein, M.T. Blackmon, and W.T. Park of General Dynamics Corporation for Materials Laboratory, Air Force Wright Aeronautical Laboratories, Wright-Patterson Air Force Base, Ohio, April 1980.

Acknowledgements

This Robotics Application Guide (RAG) was prepared by General Dynamics Fort Worth to fulfill contractual obligations of the Task A portion of Contract F33615-78-C-5188, entitled "A Robotics System for Aerospace Batch Manufacturing." The contract is sponsored by the Materials Laboratory, Air Force Wright Aeronautical Laboratories, Air Force Systems Command, Wright-Patterson AFB, Ohio. It is being administered under the Computer Integrated Manufacturing Branch, Dennis E. Wisnosky, chief. The project engineer for this ICAM project (No. 812-8) is Michael J. Moscynski. This guide was compiled at General Dynamics by Mark T. Blackmon and Lindon L. Toepperwein. The contractor was assisted in preparation of this document by a subcontracted coalition of robot experts consisting of

Mr. Ron Fukui (616) 349-8761
PRAB Conveyors, Inc.
Versatran Division
5944 E. Kilgore Road
Kalamazoo, Michigan 49003

Dr. William T. Park (415) 326-6200 Ext. 2233
SRI International
Rm. J2039
333 Ravenswood Avenue
Menlo Park, California 94025

Mr. Brian Pollard (203) 744-1800 Ext. 309
Unimation, Inc.
P. O. Box 651980
Danbury, Connecticut 06810

Mr. Don Seltzer (617) 158-1368
Charles Stark Draper Laboratories, Inc.
555 Technology Square
Cambridge, Massachusetts 02139

Mr. Ron Tarvin (513) 841-8753
Cincinnati Milacron, Inc.
4701 Marburg Avenue
Cincinnati, Ohio 45209

In addition to the subcontracted coalition, information from the automation laboratories of the National Bureau of Standards was adapted for use in the main text and glossary.

-1-

Introduction

Automated technology was developed in the arena of large-scale mass production where the working machines were designed for and dedicated to one application for their entire lifetime. This approach to automation is not always feasible in aerospace manufacturing. Items are produced in much smaller quantities or batches that do not support the high capital investment and changeover costs associated with hard automation. Until recently, the alternative was a labor-intensive approach. This solution has become more costly as wages increase and average productivity per worker decreases. The search for a third viable alternative bred the concept of programmable or flexible automation.

Flexible automation still involves relatively high capital investment, but changeover costs are significantly reduced. In this scheme, the automation is able to process part configurations with a wider range of variation than hard automation applications. Thus the low quantity per batch that makes hard automation economically infeasible in aerospace applications is compensated for by broadening the class or range of similarities that the technology is able to process. This classification of operations into groups of similar processes is termed group technology. The changeover from one job to the next is accomplished through programming rather than restructuring or replacing the hardware. In this plan, functionally unrelated parts may be grouped together because the processes required to produce them are similar and because they are manufactured at the same station. Machinery is utilized throughout its lifetime thus justifying the capital investment, and low changeover costs make this ideal for a batch-manufacturing environment. Robotics technology belongs to this class of automation.

The purpose of the Robotics Applications Guide (RAG) is to provide an introduction to new robotics technology in the aerospace community. The intent of the RAG is to introduce the concepts of robotics manufacturing technology, to provide a workable approach to implementation of this technology, to provide a means of determining costs of implementation, and to review and document currently available literature on robotics. This guide contains the sections described below.

Section 2, Robot Configuration, defines the basic characteristics of a robotic system. The information therein describes what robots are, what their capabilities are, and how they work.

In Section 3, Sensors, various kinds of sensors are discussed and classified. Application methods and use considerations are also included for a complete examination of this aspect of a robotic system.

Section 4, Tooling, provides information on four classifications of robot tooling and explains their function and design. Interfacing methods and application information are also included.

Section 5, Work Station Integration, is a guide for robotic system design. Topics of discussion are focused toward an integrated programmable robotic system for batch manufacturing and include programming, control functions, and control structures.

Section 6, Application Information, is a discussion of the factors that determine current and potential uses of robots. Subsections contain implementation procedures, safety, and guidelines for economic justification and analysis.

Two appendices are included in the guide: Appendix A, Glossary of Robotics Terms, and Appendix B, List of Current Robotic Literature.

The RAG should eliminate some of the more tedious preliminary research work necessary before any serious attempt is made at implementing new robotic technology.

-2-

Robot Configuration

The term robot is often used but seldom understood. A robot may be broadly defined as a machine which is in some way physically similar to a man or which performs a function similar to that performed by a man. The following definition highlights the features that a mechanism must have to qualify as an industrial robot. An industrial robot is a programmable multifunctional device designed both to manipulate and transport parts, tools, or specialized manufacturing implements through variable programmed paths for the performance of specific manufacturing tasks.*

Industrial robots are devices that perform tasks too physically demanding, menial, or repetitive for a man to do efficiently. Industrial robots generally consist of an arm, to which an end effector (gripper, spot welder, drill) is affixed; a power source supplying electrical or hydraulic power; and a control unit providing logical direction for the unit.

2.1 MANIPULATOR HARDWARE

Robots are manufactured in a variety of ways. Four of the parameters for specifying manipulator arms are motion, actuation, range, and capacity. These areas are discussed in the following subsections for the purpose of defining and describing the configuration of manipulator hardware.

2.1.1 Configurations

Although robots vary widely in configuration, mechanically most fall into one of four basic motion-defining categories: jointed arm, Cartesian coordinate, cylindrical coordinate, and spherical coordinate (Reference 1).

2.1.1.1 Jointed Arm

The jointed-arm robot most closely resembles a human arm. This type of arm consists of several rigid members connected together by rotary joints as shown in Figure 1a. In some robots, these members are analogous to the human upper arm, forearm, and hand; the joints are equivalent to the human shoulder, elbow, and wrist. A robot arm of this type is usually mounted on a rotary joint whose major axis is perpendicular to the robot mounting plate. This axis is known as the base or waist. Three axes are required to emulate the movements of the human wrist. These axes can be called pitch, yaw, and roll. (Refer to Appendix B, Product Literature, Cincinnati Milacron T3 Robot.)

2.1.1.2 Cartesian Coordinate

Cartesian-coordinate robots consist of orthogonal slides (prismatic joints) and a nonrotary-base axis as shown in Figure 1b. The end effector is positioned within a Cartesian-coordinate system. Some systems utilize rotary actuators to control end-effector orientation. Robots of this type are generally limited to special applications. The SIGMA robot, manufactured by the Italian company, Olivetti, is the only Cartesian robot used to any extent in industry (mainly in Olivetti plants) (Reference 2). Thermwood's Cartesian-5 machine could perhaps be considered a Cartesian-coordinate robot too.

---

*Definition from Robotics Today, published by the Society of Manufacturing Engineers, R. N. Stauffer, Editor, Dearborn, Michigan, Fall 1979.

2.1.1.3 Cylindrical Coordinate

Cylindrical-coordinate robots are constructed of a number of orthogonal slides and a rotary-base axis as shown in Figure 1c. Additional rotary axes are often used to allow for end-effector orientation. Cylindrical-coordinate robots are best applied when the tasks to be performed or machines to be serviced are located radially from the robot and no obstructions are present. The PRAB Versatran Model F600 is a good example of this type of robot.

2.1.1.4 Spherical Coordinate

Spherical-coordinate robots are similar to a tank turret; they consist of a rotary base, an elevation pivot, and a telescoping extend-and-retract boom axis as shown in Figure 1d. Up to three rotary wrist axes - pitch, yaw, and roll - may be used to control the orientation of the end effector. The Unimate 2000B is an example of a spherical-coordinate robot.

2.1.2 Actuators

Industrial robots generally use one of three types of drive systems - hydraulic, electric, and/or pneumatic.

Hydraulic robots have the advantages of mechanical simplicity (few moving parts), physical strength, and high speed. This type of robot generally uses hydraulic servo valves and analog resolver units for control and feedback. Digital encoders and well-designed sensitive feedback control systems can give hydraulically actuated robots an accuracy and repeatability usually associated with electrically actuated robots. A characteristic often thought of concerning hydraulic robots is oscillation or bounce in moving and decelerating to a point. By programming a delay, to allow for settling prior to tool function, the difficulties caused by oscillation can be eliminated.

Electrically actuated robots are almost all driven by DC stepping motors. These robots tend not to be as strong or as fast as hydraulic robots, but they do exhibit good accuracy and repeatability properties, particularly when ball-screw drives are used. Because electric robots do not require a hydraulic power unit, they save floor space and decrease factory noise.

Pneumatic drive systems are generally reserved for small limited-sequence pick-and-place applications. Techniques for servoing joints with pneumatic actuators are in research (References 3,4).

Softness and stiffness are actuator characteristics which are often referred to but seldom specified. The major concerns in most applications are accuracy and repeatability, and if the system as manufactured can meet the requirements, then degree of stiffness is largely irrelevant. There are other applications in which stiffness is a critical factor, as in the case of a robot holding a tool at a point while external forces are applied against it (such as in a drilling operation). Rigidity of this type is a variable which can be changed to suit the purpose by minor adjustment to the mechanical actuators and/or the electrical compensation devices. In order to avoid warranty problems, these adjustments should only be made with the robot manufacturer's full consent and cooperation.

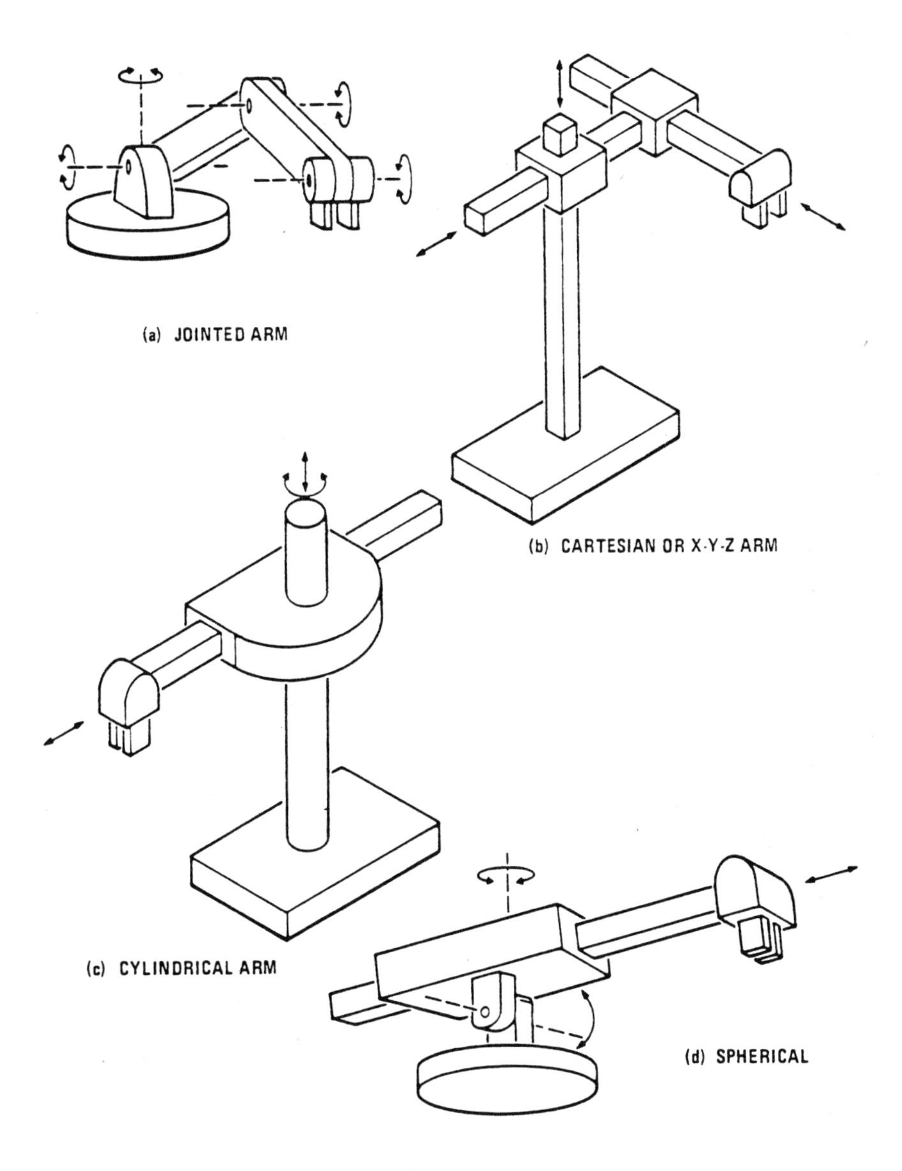

Figure 1
FOUR COMMON ARRANGEMENTS OF ROBOT MANIPULATOR JOINTS

2.1.3 Work Volume

The size and shape of the work volume is one of the most important characteristics to consider when choosing a robot for a particular application. Robot manufacturers' brochures usually describe the work volume, envelope, or range by one or more scale drawings with dimensions.

What the manufacturer means by work volume must be known. Work volume generally means the region which can be reached by some point on the wrist of the manipulator, not the tool tip. This is because the manufacturer cannot predict the shape or size of the tool which the customer may want to use. The customer must decide whether the manipulator plus the tool will be able to reach all the positions required. Generally speaking, the robot will be able to reach outside of its work volume with the tool. This extra reach should be taken into account when planning for the safety of the people working near the robot and when considering the placement of equipment around it.

The length of the tool can have some subtle effects on the effective work volume when tool orientation is taken into account. For example, a manipulator can put its wrist at some given position with a certain orientation. When a tool is mounted on the wrist, the tool tip may not be able to reach that same position and orientation in space. In an attempt to bring the tool tip to this position and orientation, one of the manipulator joints could jam against a limit stop before the position is attained. The inherent positional characteristics of the manipulator arm will also affect the work volume. Currently, for example, no six-jointed arm is able to position its wrist to any arbitrary orientation about any fixed point in its work volume. (However, an arm with a three-axis gimbal at its wrist could do so.) A manipulator with fewer than six joints is even more restricted in the placement of its wrist or a tool attached to the wrist. Therefore, the required tool orientation should be determined for each tool position in a task, and the manipulator under consideration should be checked carefully to ensure that the manipulator has the capability of attaining this orientation.

Types of manipulator joint motion are sometimes described by a shorthand method to aid in defining a robot's characteristics related to work volume. This shorthand establishes robot configuration as a classification based on the types and number of joints that make up the machine. Joints can be sliding or prismatic, designated S and P respectively, and rotary, designated R. The shorthand classification then describes a particular configuration from the base to the end-effector attach point. Therefore, an arm with three orthogonal sliding joints would be called an S3 arm. A machine with a rotary base, two sliding joints, and a rotary joint would be an RS2R arm. Note that this classification method is for arm configuration and should not be confused with end effector or robot manufacturing and model notation.

2.1.3.1 Common Work Volume Shape

The length of the links in a manipulator and the arrangement of its joints determine the shape of the manipulator's work volume. Some common work-volume shapes and joint arrangements are

- Rectangular - three orthogonal sliding joints (X-Y-Z, Cartesian, or overhead crane manipulators). IBM's manipulator is an example. (Figure 2a)

- o Cylindrical - a horizontal sliding joint that rides up and down a column and pivots left or right around it. The Versatran has an approximate cylindrical geometry (not exactly, because its two sliding joints' axes do not intersect). (Figure 2b)
- o Spherical - a sliding joint mounted on a trunnion. The Unimate has an approximate spherical geometry (not exactly, because the axis of its sliding joint does not intersect the axis of its second rotary joint). (Figure 2c)

Some manipulators, such as the IRb-60 and IRb-6, the Cincinnati Milacron T3, and Unimate 250 and 500, have rather irregular-shaped work volumes as shown in Figure 3.

2.1.3.2 Limits On Work Volume

The length of the arm, joint arrangement, and range of motion of the joints determine the limits on the work volume. Some manipulators may not be able to reach those limits if they are carrying a particularly heavy load. In this case, the limits would change and would result in a smaller work volume.

2.1.3.3 Optimization

The work volume of a robot can be improved by using various modification methods. One method of extending the work volume is to mount a long tool on the manipulator's wrist. Excessively long tools will certainly degrade the spatial resolution and may also degrade the dynamic performance due to increased inertial loads. The tool itself may be capable of making some motions.

A second, more expensive method is to mount the entire manipulator on a movable base. The base usually rolls on tracks because a manipulator is often quite heavy. The additional motion must be controlled as precisely as that of the other joints in order to preserve the overall accuracy of the system (assuming the robot is not equipped with external sensors).

In some regions of its work volume, a manipulator may be capable of much better performance than indicated on its specification sheet. The work station may possibly be arranged to take advantage of this better performance. Accuracy, repeatability, load-handling ability, and dynamics can vary from one location to another in the workspace. These areas are discussed in 2.1.4 and 2.3.

2.1.4 Load Handling Capacity

Except for arms with an X-Y-Z geometry, most arms are able to lift more weight at some locations in the workspace than at other locations. The lever arm between a rotary actuator and the load can vary with the instantaneous arm posture. Some arms can be mounted in various positions -- even upside down. However, they will usually be able to lift more when mounted in certain postions.

The manipulator must be able to carry not only the workpiece but also the gripper that holds it. Grippers are frequently much heavier than what they are designed to carry. This is especially true of grippers (and other tools) that contain their own actuators.

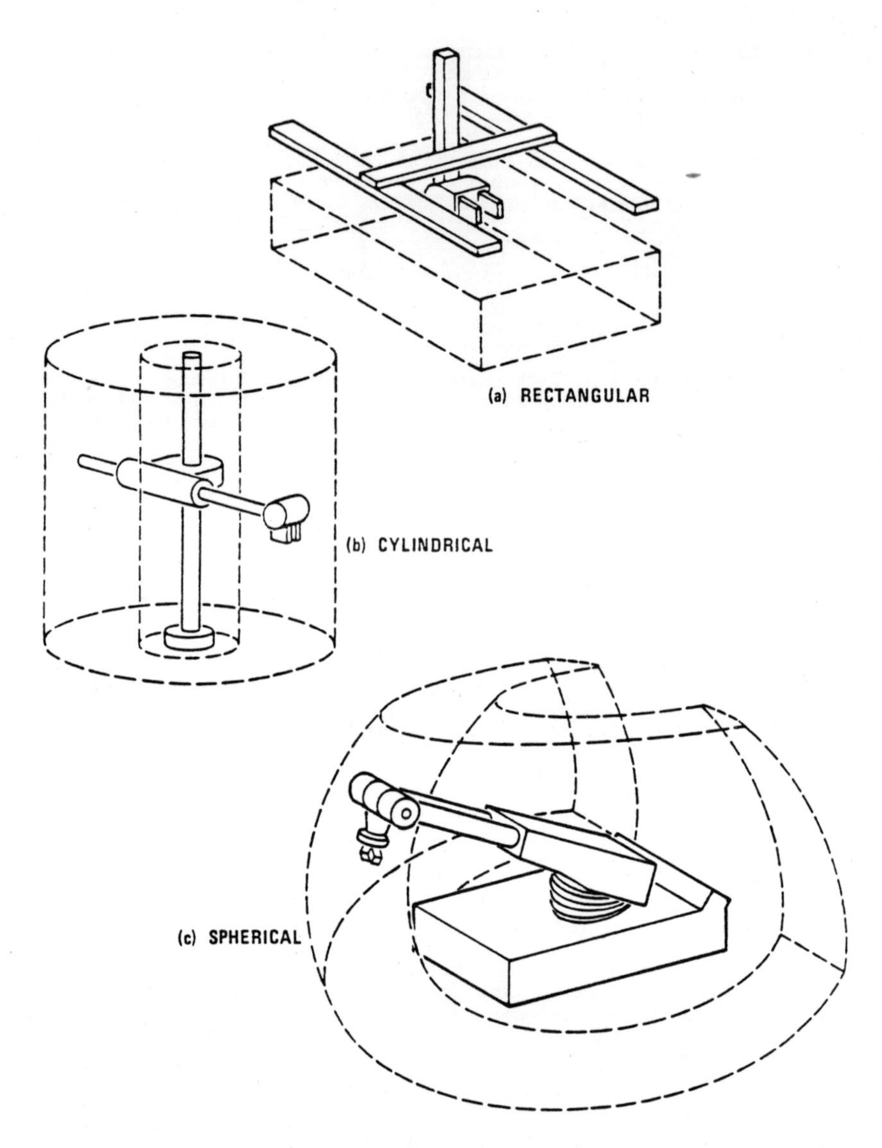

Figure 2
SOME COMMON WORK VOLUME SHAPES

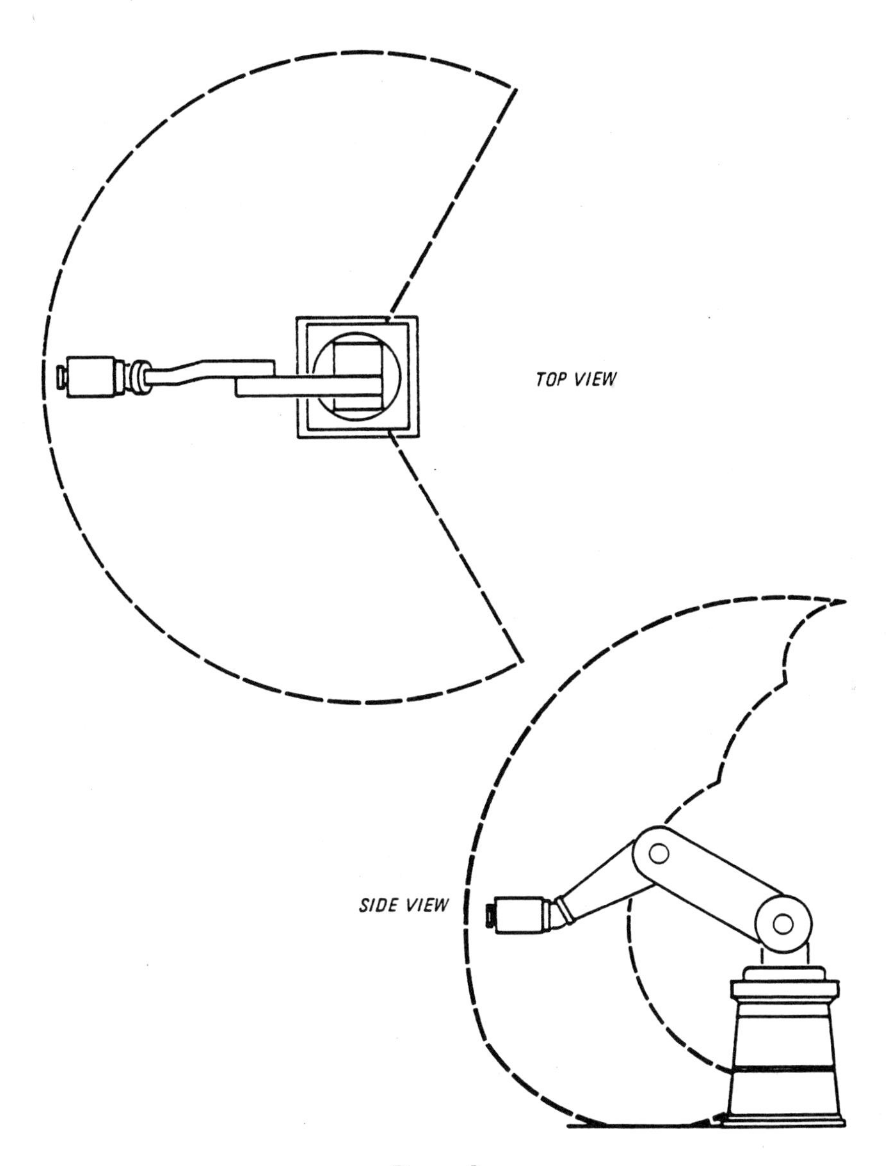

Figure 3
IRREGULAR WORK VOLUME

A robot manufacturer commonly quotes a reduced maximum velocity for loads over a certain weight to accommodate momentum.

2.2 ROBOT CONTROLLERS

The sophistication required of a robot control system varies directly with the complexity of the task to be performed. Limited-sequence robots use pneumatic, mechanical, or simple electrical logic to control motion. These systems make use of open-loop motion control. An open-loop control system is one in which the robot motion is controlled by mechanical stops and is not fed back to the controller.

In sophisticated industrial robots, the position of each joint or axis of motion is controlled by a closed-loop servo system. A closed-loop system is one in which robot axis position is measured and compared to a set point from the robot controller. If the position is different from that called for by the set point, the control system will cause the joint actuator (electric or hydraulic) to move the joint to the correct position. The robot controller generates one set of points for each axis. Each such group of set points will move the robot's end effector to a different position in the workspace.

The most sophisticated, and therefore versatile, type of robot controller is a minicomputer-based computer numerical control. This type of control is capable of providing the axis transformation required to convert "real world" (Cartesian, cylindrical, spherical) coordinate position data into robot joint-position information. It can also provide numerous other useful features, such as teach-mode part-program generation, external program storage, sensor (tactile, visible) interaction, tool center-point programming, and sophisticated program-flow modification capabilities. Less sophisticated robot controllers are available. These include simple controllers similar to those used on standard numerical control machines. These less sophisticated controllers require tedious hand programming or the support of external computers.

High-level system controls utilizing hierarchical control structures are discussed in Section 5.3.

2.3 DYNAMIC PROPERTIES

The dynamic properties of a given manipulator includes its accuracy, repeatability, stability, and compliance. These characteristics depend upon the tool and its function, the arm geometry, the accuracy of the individual point servos, and the quality of the computer programs which perform kinematic computations.

2.3.1 Dynamic Performance

The dynamic performance of a manipulator describes how fast it can move, how quickly it can stop at a given point within a certain accuracy, and how much it overshoots a stopping position. When the tool is being moved rapidly toward an object, any overshoot can be disastrous. On the other hand, moving too slowly can waste excessive amounts of time.

Good dynamic performance is usually extremely difficult to achieve in a manipulator that has rotary joints at its base. The inertial load seen by a servo controlling one of those joints depends not only on the inertia of the object being carried but also upon the instantaneous position and motion of the joints (Reference 5). The mass and moments of inertia of the rigid links in a manipulator also impose a large fraction of the total load on those joint servos during a rapid motion. An increase of ten-to-one in the inertial load on the first rotary joint with a change in posture is not unusual in a commercial manipulator. If the individual joint servos are classical proportional-integral-derivative controllers, they must be tuned for maximum inertial loads to guarantee that they will never overshoot their targets (References 6-8). This tuning seriously degrades their performance from what it might be. Much research is currently under way on advanced servo designs for manipulators (References 5,9-21).

As an example of the effect of reflected link inertia on performance,the load on the vertical rotary joint of a Unimate or Versatran is smallest when its boom is pulled in and largest when the boom is fully extended. Consider a movement in which the tool must swing rapidly from a position on the robot's left to a postion on its right at the same distance from the central rotary axis. If this movement is trained as two positions, one at each end of the trajectory, the boom of either a Unimate or a Versatran will remain extended throughout the entire motion, and the tool tip will travel in a wide arc. The arm's moment of inertia about its rotary axis will be high, the acceleration and deceleration will be small, and the total transit time will be long.

Among robot trainers, a well-known trick for speeding up such a motion is to train one or more extra via points located to bring the arm into a lower-inertia posture for part of the motion. A via point is one through which the tool tip should pass without stopping and is illustrated in Figure 4.

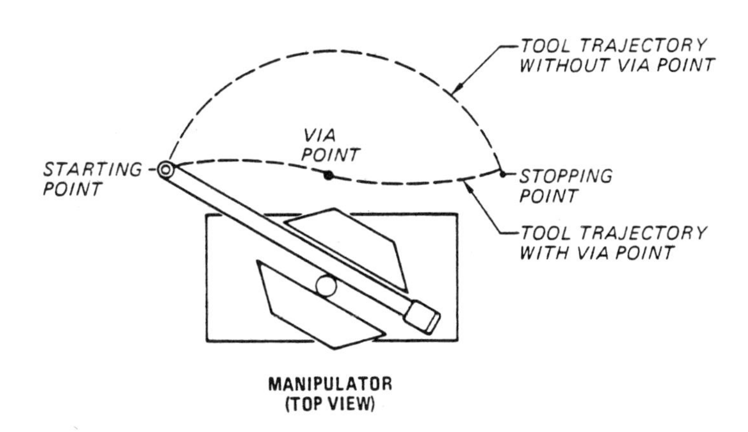

Figure 4
USE OF VIA POINTS TO SPEED UP MANIPULATOR MOTIONS

For example, one via point might be trained midway along the straight line running form the starting position to the stopping position. This procedure will force the boom of either machine to retract as it starts to move, will reduce the moment of inertia felt by the rotary joint's servo, will potentially result in larger acceleration and deceleration, and will reduce transit time. We say potentially because the servos on most arms are rather exotic nonlinear devices and because making generalizations about their performance is difficult.

Via points should be used with caution, for they can damage some arms if they are placed inappropriately. Generally speaking, no major joint should approach its position for the via point at full speed and leave it at full speed in the opposite direction. At least one arm manufacturer cautions customers that the hydraulic hoses can rupture in such a situation.

2.3.2 Stability

Stability refers to the lack of oscillations in the motion of the tool. Oscillations are bad for several reasons:

- They impose additional wear on the mechanical and hydraulic parts of the arm.
- They make the tool follow a different path in space during successive repetitions of the same movement, requiring more distance between the intended trajectory and surrounding objects.
- They can increase the time needed for the tool to stop at a precise position.
- They may cause the tool to overshoot the intended stopping position and make it collide with something.

Two different kinds of oscillations are damped and undamped. Damped oscillations are those which die out with time (transient oscillations). Undamped oscillations may persist or may grow in magnitude (unstable or runaway oscillations). Of these three types, undamped oscillations are the most serious for they can do tremendous damage to surroundings. Persistent oscillations are a borderline case; they are only observed because a manipulator as a dynamic system is highly nonlinear. Limit cycles can develop and result in steady-state oscillation. Damped oscillations are less likely to do damage but are no more acceptable.

The servo designer is to ensure that the arm never breaks into oscillation. The variation of inertial and gravitational loads on the individual joint servos as the arm's posture changes makes this difficult. Furthermore, the servos must operate over a wide dynamic range of position error (and in some cases, of velocity error), and they must work reliably in all situations despite the limits on velocity and acceleration imposed by the actuators used.

One robot controller locks each joint independently the first time it reaches its set point. Special circuitry also decelerates the joint after it comes within a certain distance of that position. The joints in this robot may lock in any order. When the joints are all locked (a condition called total coincidence), the arm is stationary, and it can then begin moving to the next position. If the position is held for more than a few seconds, the tool slowly creeps away from its programmed position as oil leaks out of the actuator cylinders. When the position error accumulates sufficiently, the joint servos are allowed to operate again to return the tool to the original position. This is technically a form of instability in the sense that the tool position can vary periodically (although the period may be on the order of 30 to 60 seconds). However, it is part of the machine's normal operation and causes no problems.

Another robot manufacturer allows the joint servos to operate continuously. Sophisticated servo designs derived from experience in building NC tools prevent oscillations from starting regardless of the load carried.

Certain exceptional conditions can be extremely unstabilizing to a joint servo system. A classic example is what happens when the load accidentally slips out of the end effector. This causes a step change in the gravity loading on one or more joints and can cause a poorly designed arm to go into oscillation. Motion of a joint can also exert various combinations of inertial, centrifugal, and Coriolis forces on the other joints. The reactions of the other joints to these forces can exert forces on the original joint, and this is another potential source of oscillation. Finally, two manipulators working in close proximity can excite oscillations in each other. This can either be through a mechanical coupling such as a common mounting or through a workpiece held simultaneously by the two machines.

2.3.3 Spatial Resolution

Spatial resolution is a descriptive element of the movement of a robot at the tool tip. Resolution is a function of the design of the robot control system and specifies the smallest increment of motion by which the system can divide the working space. This may be a function of the smallest increment in position the control can command, or it may be the smallest incremental change in position that the control measurement system can distinguish. Spatial resolution is the control resolution combined with mechanical inaccuracy. In order to determine spatial resolution, the range of each joint on the manipulator is divided by the number of control increments. For example, Figure 5 describes a 48-inch sliding joint and a control system using 12-bit storage for a capacity of 4096 command increments. The control resolution for this system is 0.012-inch (0.30 mm). The spatial resolution then is the control resolution plus mechanical inaccuracies. Mechanical inaccuracy is discussed further in the next section.

Two manipulator positions that differ by only one increment of a single joint are called adjacent. A unit change in the position of a sliding joint will move the tool tip the same distance, regardless of where it is in the workspace. A manipulator with an X-Y-Z geometry therefore has essentially constant spatial resolution throughout its work volume. This consideration could be important if the arm is to be trained to perform a precise manipulation in one location of its workspace and then is to repeat it elsewhere in the workspace.

However, a unit change in the position of a rotary joint will move the tool tip through a distance that is proportional to the perpendicular distance from the joint axis to the tool tip. For example, some manipulators have a rotary joint with a vertical axis that carries all the other joints and links. The servo on this joint can reliably position the boom of the manipulator to a given orientation about this vertical axis within a certain maximum error. The effect of this angular-position error on the final tool-tip position obviously depends upon how far the boom is extended. The farther the boom is extended, the larger the distance that the tool tip will move when the rotary joint moves to an adjacent position as shown in Figure 6.

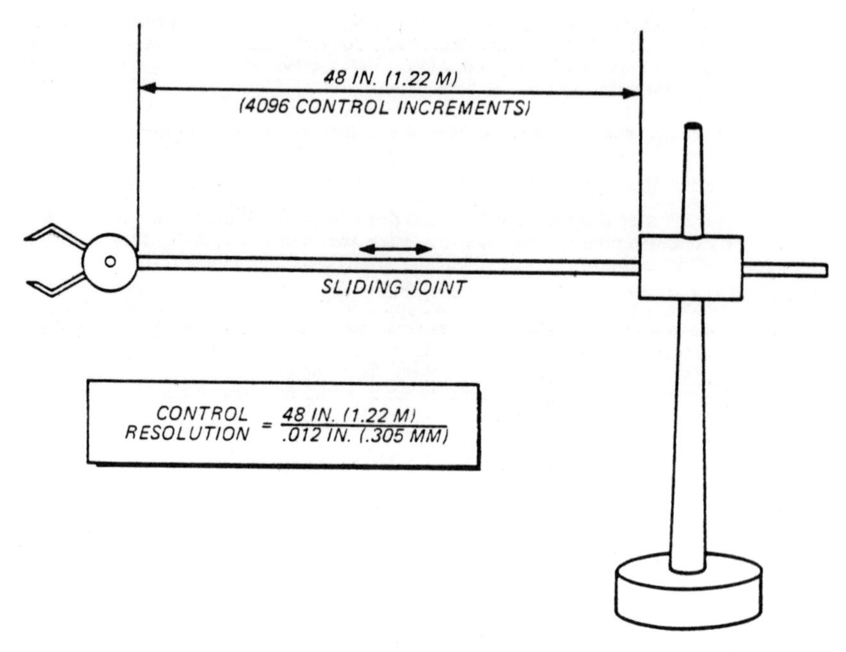

Figure 5
CONTROL INFLUENCE ON RESOLUTION

A long tool can make precise training very difficult by reducing the spatial resolution of the wrist joints. A unit increment in the position of a wrist joint could easily displace the tool tip much farther than a unit increment in the position of a nonwrist joint. A contributing factor is that the manipulator manufacturer may put a lower-resolution position feedback transducer in a wrist joint since its lever arm will be only the tool length, not the arm length.

When training a manipulator that has no computer, the trainer usually has to use a button box (teach gun or pendant) which can only move individual joints at fixed rates. Consequently, the trainer tends to make many small motions, moving one joint at a time, until he gets the tool tip exactly where he wants it. Then, when he attempts to correct the tool orientation by moving a wrist joint, the tool tip swings away from where it should be. A computer can eliminate much of this annoyance for the trainer, as in the Cincinnati Milacron T3's tool center-point control mode for example, that translates tool point motion into the joint motion needed to move the tool point as desired. Training a precise positioning task without help from a computer is easier if the spatial resolution of the wrist joints with the given tool is better than that of the other joints.

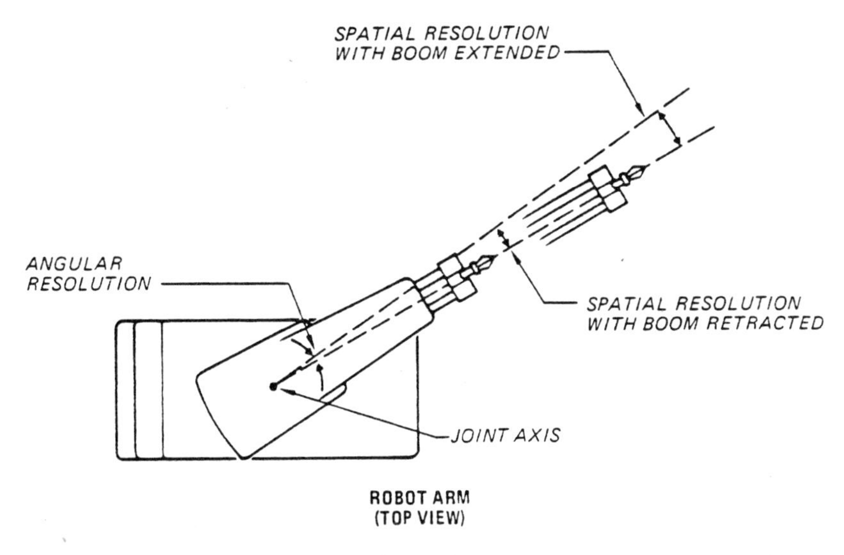

ROBOT ARM
(TOP VIEW)

Figure 6
EFFECT OF BOOM EXTENSION ON SPATIAL RESOLUTION
(EXAGGERATED)

2.3.4 Accuracy

Accuracy is a term often confused with resolution and repeatability. Three factors are brought together to describe the characteristic or specification known as accuracy as related to robots. The three factors are

1. The resolution of the control components
2. The inaccuracies of the mechanical components (linkages, gears, beam deflection, etc.)
3. An arbitrary never-before-approached fixed position (target).

For explanation, consider a single-joint machine with negligible mechanical inaccuracy and a control resolution of 0.012-inch (.305 mm). The accuracy with which this machine can approach an arbitrary target is one-half the distance between two adjacent control positions or 0.006-inch (.152 mm), as depicted in Figure 7.

When the inaccuracies associated with the mechanical components are included, a poorer accuracy will result. As shown in Figure 8, the inaccuracies that contribute to the largest positional error, establishing the worst condition, are used to determine a realistic spatial resolution from which accuracy is derived. Some of the factors that

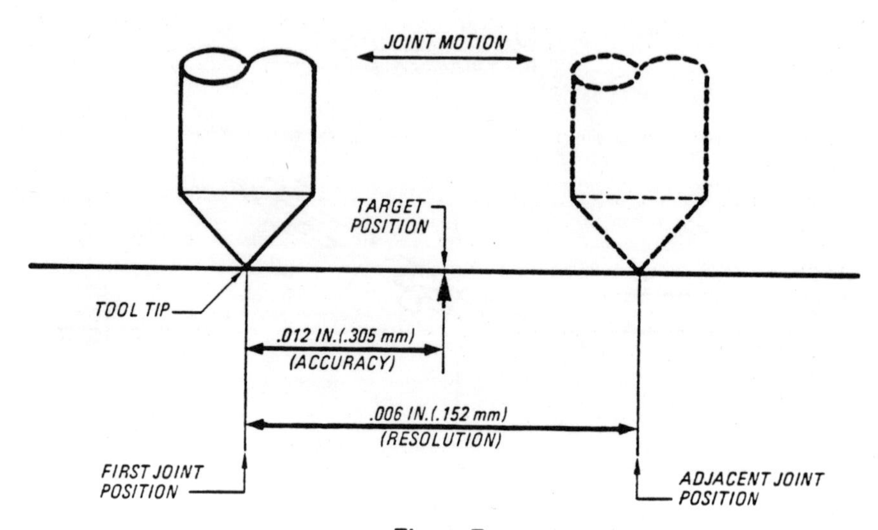

Figure 7
ACCURACY/RESOLUTION RELATIONSHIP

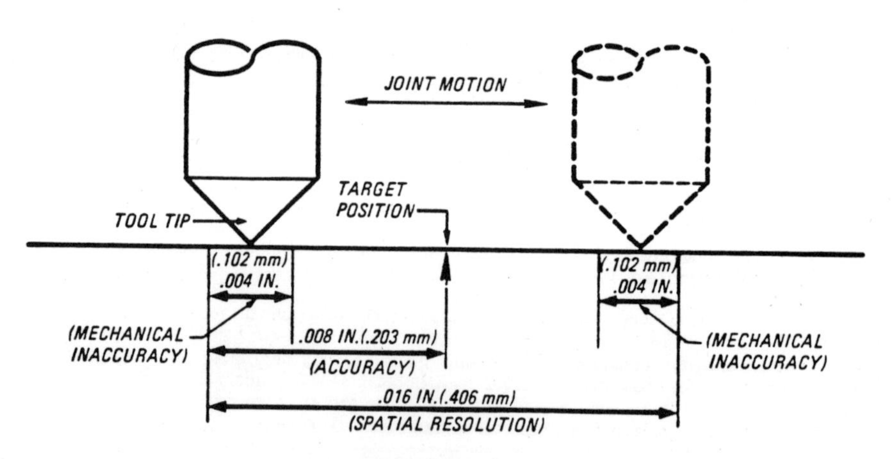

Figure 8
ACCURACY/SPATIAL RESOLUTION RELATIONSHIP

contribute to these inaccuracies are backlash in the gears, loose linkage, and the effects of the payload to be handled. Backlash has more influence in rotary axes where the feedback element is located at the rotational joint and the payload extends some distance away from the axis. At high payload weights, beam deflection will begin to affect and reduce accuracy. Beam deflection exists in gravity-effected axes (axes near a horizontal orientation) under static conditions and in all axes under dynamic conditions. Beam deflection can also lead to severe resonant oscillation if drive backlash is present.

In a sense, it is meaningless to speak of the accuracy of a robot that is operated only in a "tape-recorder" mode. In that mode, the control system merely records joint positions during training and plays them back later during production. In such applications, repeatability and resolution are the significant performance specifications. The resolution specification determines whether the manipulator can reach positions closely enough to do the job the first time during training. The repeatability specification determines whether it will be able to reach them closely enough to do the job the second and succeeding times during production.

Accuracy is only meaningful in describing a robot manipulator in which a computer in the control system has to calculate a set of joint positions that will place the tool tip in a position that is described in some manipulator-independent coordinate system. Such calculations are necessary in manufacturing situations in which:

- The tool used during training is not the same size and shape as the one that will be used during production
- A sequence of operations is trained either on a stationary object and performed or on a object that is moving or in a different position
- Robot motions are computed from geometric information about work-piece dimensions.

In such situations, infinite resolution and perfect repeatability are of no use if the kinematic calculations are inaccurate, because each position is calculated under changing or new conditions, and depends completely on the control system calculations.

When the robot's position is calculated as in off-line programming, another aspect of accuracy is important - the correspondence between actual measurement and control system measurement. Perhaps the following example better relates this concept of accuracy. Assume the robot is commanded to move 20 inches (50.8 cm), and the actual move is measured and found to be 19.90 inches (50.55 cm). The error is 0.10 of an inch (.25 cm) and can be represented as an accuracy error of 0.5 percent less than the commanded distance. If by test the error is consistent over the range of the robot, the situation can be remedied simply by scaling all movements to account for the error. If the error is not linear over the entire range, then other means of adjustment within the control itself may be necessary. The accuracy error illustrated here can have several causes but are usually due to numerical error in the computation of the joint positions or to an inaccurate reference measurement.

Accuracy of the robot can be discussed relative to global accuracy and local accuracy. Global accuracy refers to the accuracy of any point within the working range of the robot. Local accuracy refers to the accuracy of a point in the neighborhood of a zero reference point within the working range. Local accuracy may be more significant since position points are generally programmed from a reference point.

2.3.5 Repeatability

Repeatability is the ability of the robot to reposition itself to a position to which it was previously commanded or trained. Repeatability and accuracy are similar; however, they define slightly different performance concepts. The three factors used in describing accuracy in the previous subsection can be modified to explain repeatability. Briefly, the three factors are resolution, inaccuracy of components, and an arbitrary target position. Repeatability is affected by resolution and component inaccuracy; however, it is not relevant to an arbitrary target position. When speaking of repeatability, only the ability of the machine to return to a previously trained position is considered. By the definition of accuracy (one half the distance between two adjacent positions nearest an arbitrary target) and since the arbitrary position is eliminated and replaced by the previously taught (best resolved) position, the repeatability will always be better than the accuracy if other influences discussed later are minimized.

Figure 9 is a simple example of repeatability. Initially, the robot, limited by control resolution, is positioned as close to the arbitrary target as possible. This places the robot at position T. The robot is then moved away and commanded to return automatically to position T. When the robot attempts to return to the previously taught position, inaccuracies within the control system and mechanical components allow the robot to stop at position R. The difference between position T and position R is a measure of the repeatability of the robot.

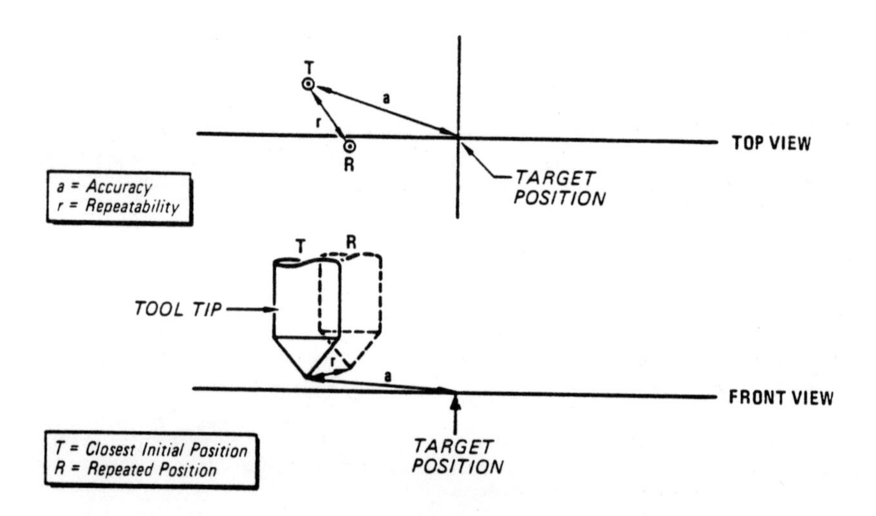

Figure 9
ACCURACY – REPEATABILITY RELATIONSHIP (EXAGGERATED) SHOWING REPEATABILITY BETTER THAN ACCURACY

Both short-term and long-term repeatability exist. Long-term repeatability is of concern for robot applications requiring the same identical task to be performed over several months. Over a long time period, the effect of component wear and aging on repeatability must be considered. For many applications where the robot is frequently reprogrammed for new tasks, only short-term repeatability is important. Short-term repeatability is influenced most by temperature changes within the control and the environment, as well as transient conditions between shutdown and startup of the system. The factors that influence both short-term and long-term repeatability are commonly referred to as drift.

A review of spatial resolution, accuracy, and repeatability provides the following relationships shown in Figure 10.

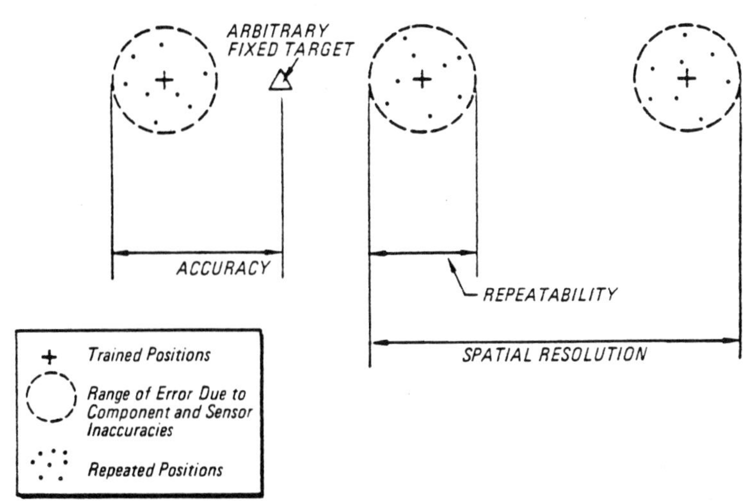

Figure 10
A TWO-DIMENSIONAL DEPICTION OF TOOL TIP POSITIONS OF ADJACENT INCREMENTS, TRAINED & REPEATED (EXAGGERATED)

- Spatial resolution describes the smallest increment of motion at the tool tip that the robot can control.
- Accuracy relates the robot's spatial-resolution-defined positional ability (including mechanical inaccuracies) to an arbitrary fixed-target position.
- Repeatability describes the positional error of the tool tip when it is automatically returned to a position previously taught.

- o Repeatability will generally always be better-than accuracy exclusive of drift.

Obtaining good repeatability is more difficult in a computer-controlled manipulator that records tool positions rather than joint positions because three additional data processing steps are involved. These three steps, which can introduce positioning errors, are

1. Converting the several joint positions to a tool position and storing it. This is called the back solution.

2. Transforming a tool position in some useful way such as by translating, rotating, or scaling it. (This step is unnecessary in simple record-playback applications.)

3. Converting the transformed tool position back to a set of joint positions. This is called the arm solution.

The way in which the computer performs the three computations above can have a profound effect upon the accuracy and repeatability of the manipulator. The accuracy of each of these operations depends upon the number of bits of precision used to store each of the representations and upon the accuracy of the algorithms used in any computations, such as taking square roots and evaluating trigonometric functions. Generally speaking, the more bits carried, the better the numerical accuracy is. However, it is possible to lose much or all of the precision in poorly coded computational algorithms. Some practical systems use floating-point representations, and others use scaled-integer representations. Round-off errors should be given careful attention in all cases, while overflow and underflow must be prevented in scaled-integer computations.

A requirement that has not received much attention thus far is that the kinematic equations used in the arm solution and back solution must accurately reflect the design of the manipulator. The accuracy of these computations depends upon the accuracy with which the following four values (joint parameters) are known:

a. the joint extensions and rotations
b. the link lengths
c. the offset distances between successive joint axes
d. the angles between successive joint axes.

The values mentioned in (a) are usually accurately known. In some arms, these values are obtained by scaling and offsetting a value read from a precise displacement or rotation transducer, such as linear variable-differential transformer (LVDT), a resolver, or an optical digital encoder. The calibration factors for such a transducer may be measured easily and generally do not change.

The values mentioned in (b), (c), and (d) should ideally be obtainable from the blueprints for the manipulator itself. In more precise arms, thermal expansion of the links may become important, but violent collisions are not likely to deform an industrial arm significantly. If the manipulator manufacturer intends to use an arm controller that only records and plays back joint positions, there is no compelling reason for him to control the arm dimensions closely during manufacture. If these quantities only vary slightly from one arm to another, then it will still be practical for his customers to use the joint position data from one manipulator to operate other manipulators with only minor touch-ups at critical steps. This is satisfactory for most applications today.

If such an arm is retrofitted with a computer controller for this purpose, then these quantities should be measured accurately. Their values should then be incorporated into the computer code that performs calculations (1) and (3) previously mentioned.

In most arms, computation (3), converting a tool position to a set of joint positions, is the most difficult to accomplish exactly. The programmer of the manipulator-control computer usually assumes that the angles between successive joint axes --joint parameter (d) above -- are all multiples of 90 degrees. This makes many terms with sine and cosine factors drop out of the kinematic equations (References 21-23). The computation can then be performed in a very short time with relatively little code. Unfortunately, most manipulator manufacturers have no reason to align these joint axes very precisely because they expect users to record and play back only joint positions. If the axes are slightly skewed, then for the robotic system to be able to go accurately to a point in space specified by a set of Cartesian coordinates, its computer must perform more extensive computations.

The manipulator's geometry and instantaneous posture can also lead to large position errors. One way in which this can occur is when two rotary-joint axes become closely parallel at one point along the tool's trajectory. As the axes become more closely aligned, a small change of tool orientation in certain directions will require larger and larger changes in some of the joint positions. This is called a singularity in the kinematic equations of the arm. The direction of orientation change that will cause the problem is usually a rotation of the tool about an axis normal to the plane of the aligned axes. This problem is similar to the condition of gimbal lock that can occur in a gyroscope mount, and the problem could be avoided in the same way as it is avoided in some gyroscopes --by adding another joint. Mathematically, this condition is signaled by the Jacobean matrix of the manipulator (considered as a linkage) becoming singular (having no inverse) (Reference 24). The Jacobean tells how much the tool will move or turn in any direction per unit of motion of any joint. The inverse of this matrix describes the joint motion required per unit of tool motion or rotation in any given direction. Certain terms in the inverse will tend to infinity as the axes become aligned.

Erratic motion can result in the neighborhood of a singularity due to slight errors introduced by such causes as round off. If a succession of tool positions passing near a singularity were specified with complete accuracy, then smooth motion would result. Some of the joints would have to move very rapidly for a short while, but their motions would be smooth. The effect of computation errors is to perturb the successive tool positions slightly but randomly from their ideal values. For the reasons described above, these small random perturbations in the tool motion are amplified into large random perturbations in the individual joint motions near the singularity. The rapid accelerations and decelerations of those joints as they try to follow their wildly varying set points will cause jerky erratic tool motion.

The usual format for recording tool position in the workspace is a set of Cartesian (X-Y-Z) coordinate values that represent the position of the tool tip and a set of three angles that represent the orientation of the tool body. There is no general agreement on the best set of angles to use to describe the orientation. Pitch, yaw, and roll about the tool's main axis (if it has one) is one choice. The combined position and orientation information is often referred to as the Cartesian position. Because this representation is manipulator-independent, it is simple to transform in useful ways to increase the versatility of the robotic system. For example,

- Drift during normal operation can be compensated for by periodically locating (with a sensor) three points located in known positions with respect to the workpiece. This locating gives sufficient information to correct for drift in translation and orientation. When the locating is performed at the beginning of a task, it makes precise jigging fixtures unnecessary.

- A sequence of operations can be trained on a workpiece in one position and then performed on other workpieces located in different positions and orientations. This is very useful when the robot has to process an array of workpieces on a rack or in a bin or when it is desirable to overlap the setting up of one workpiece with the processing of a second.

- A sequence of operations can be trained on a stationary object and then performed later on a object moving along a conveyor belt. The X-Y-Z positions recorded during training need only to be transformed by adding the instantaneous X-Y-Z position of the object during playback.

- In a master-slave teleoperator mode of operation, the Cartesian position of the master can be computed from its joint postions, and then the joint positions of the slave arm can be computed from the Cartesian positions. This allows the master arm to have a different geometry and scale from that of the slave arm, i.e., to make the slave arm more convenient to carry around or to operate in cramped quarters. No computer-controlled industrial manipulators currently offer this useful type of control, unfortunately.

- Manual control can be made much easier for the operator by translating signals from a button box into smooth motion of the tool tip along a straight line in space or into rotation around the tool tip. (Cincinnati Milacron calls this "Tool Center Point Control.")

2.3.6 Compliance

The compliance of a manipulator is indicated by its displacement relative to a fixed frame in response to a force (torque) exerted on it. The force (torque) may be a reaction force (torque) that arises when the manipulator pushes (twists) the tool against an object, or it may be the result of the object pushing (twisting) the tool. High compliance means the tool moves a lot in response to a small force, and the manipulator is then said to be spongy or springy. If it moves very little, the compliance is low and the manipulator is said to be stiff.

Compliance is a complicated quantity to measure properly. Ideally, one would find the relationship between disturbances and displacements to be linear (displacement or rotation proportional to force or torque), isotropic (independent of the direction of the applied force), and diagonalized (displacement or rotation occurring only in the same direction as the force or torque), constant with time, and independent of tool position, orientation, and velocity.

In practice, a manipulator's compliance turns out to be none of these. It is a nonlinear, anisotropic, tensor quantity that varies with time and with the manipulator's posture and motion. It is a tensor because a force in one direction can result in displacements in other directions and even rotations. A torque can result in rotation about any axis and displacement in any direction. A six-by-six matrix is a convenient representation for a compliance tensor. Time can affect compliance through changes in the temperature, and hence viscosity, of hydraulic fluid for example.

Furthermore, the compliance will often be found to be a function of the frequency of the applied force or torque. A manipulator may, for example, be very compliant at frequencies around 2 Hz but very stiff in response to slower disturbances.

Finally, the compliance may exhibit hysteresis. For example, the servos in at least one hydraulic manipulator turn off when the arm stops moving. In this condition, the servo valves are all closed, and the compliance has a value that is determined by the volume of incompressible hydraulic fluid trapped in the hydraulic hoses and the elasticity of those hoses. However, if an outside force on the tool should move any of the joints more than a certain distance from the position at which they are supposed to remain, then the servos on all the joints will turn on again. The compliance then changes to a completely different value (presumably stiffer in some sense).

Electric and hydraulic manipulators both have complicated compliance properties. In an electric manipulator, the motors generally connect to the joints through some kind of mechanical coupling such as a leadscrew, pulley block, spur gears, or harmonic drive. This is because electric motors generally produce much less force or torque than a hydraulic actuator of the same size, so they require a mechanical impedance matcher between them and the joint if they are to overcome the loads that are encountered in a typical manipulator. A hydraulic actuator, however, can usually drive a joint directly.

The sticking and sliding friction in such a coupling and in the motor itself can have all sorts of strange effects on the compliance measured at the tool tip.

In particular, some of these couplings are not very back-drivable. For example, if you push on the nut of a leadscrew (back-drive), the leadscrew will not turn (unless the screw's pitch is very coarse and ball bearings are used between the threads to reduce friction). But you can turn the screw easily, and the nut will move.

Thus, a coupling that is not back-drivable actually acts like a brake that is applied whenever the servo is off. If an application requires a robot to position a tool precisely and then hold it there while it exerts a large force on a workpiece, then such a coupling can be very beneficial. Drilling would seem to be a good example of such an application. On the other hand, routing would probably not be improved by having such couplings in a manipulator, because the joints would be moving most of the time and the braking effect would largely disappear.

The friction in the coupling in a joint servo generally decreases once the joint starts moving. It can decrease so much that a force on the tool tip can now affect the tool's motion - the brake has been released so to speak. Therefore, the compliance of a manipulator with such couplings in its servos can be vastly different, depending upon whether you measure it when the tool is stationary or when it is moving. As noted above, hydraulic manipulators generally don't need couplings to provide a mechanical advantage for their actuators, so they don't have a built-in automatic braking mechanism. This may be one reason why some feel that an electric servo is inherently stiffer than a hydraulic one.

On the other hand, the fact that hydraulic fluid is incompressible leads others to think that hydraulic servos must be stiffer than electric ones. The compressible hoses in a hydraulic system combined with a long lever arm from the base of the manipulator to the tool can produce a lot of compliance. The effective stiffness of a hydraulic arm actually depends upon how all the components of the system work together. When the tool is stationary, all the servo valves will be nearly closed (some will be open a little to supply leakage flow through the largest, gravity-loaded actuators). When the tool is moving fast, as in spray-painting for example, some of the valves are open a fair amount, and the model of a fixed volume of incompressible fluid trapped in an actuator is no longer valid. The compliance will then be determined by the overall dynamic behavior of each servo loop.

Most manipulators are operated open-loop in the sense that they go blindly to a given point in space without regard to the actual position of the object in the environment or to any reaction forces (feedback) that those objects exert on the arm (or tool). In this case, less compliance than that of surrounding objects is advantageous because it means contact with objects would cause high-frequency oscillations which can be filtered out without degrading overall response. Such filtering actually requires no special effort since the combination of servo valves and actuators commonly used have relatively low bandwidths (perhaps one or two Hz).

Sensors that measure forces and moments exerted on the tool can allow the manipulator to track or locate objects by touch. However, oscillations may arise in the force-feedback control loop if the compliance at the point of sensing is too low (too stiff). Our familiarity with the dynamic behavior of the limbs of our own bodies is extremely misleading in predicting the performance that might be expected if a mechanical arm is equipped with touch sensors or force and moment sensors. This familiarity misleads in the following ways:

- Our limbs can be made either very stiff or very flexible as the situation demands. The fingers provide an extra measure of compliance with low mass so that we can search quickly, yet we are able to stop our gross motion before the bulk of our limb collides with an object. Imagine a blind man searching for the exit in a china shop by using only his elbows to feel. This situation arises when we mount a stiff sensor on a stiff robot arm.
- The tremendous help we get from our eyes in estimating the location of an object when we reach for it is often ignored. (Imagine reaching for an object when there is an invisible pane of glass in the way.) We precompute the gross motion and only use our sense of touch in the last inch or so of travel. When performing a very familiar task, we may not look for an object if we expect it to be in a certain place, but we will use our memory of where it should be in order to throw our hand at it.
- Our hands really do bump into a great many things as we perform tasks, and no damage results. In part, this is because most things are a great deal stronger than flesh. A large hydraulic-powed manipulator working with sheet metal parts has to be as careful as, in human terms, a blind glassblower.

The conclusion is that one has to examine the particulars of a given servo design in order to predict whether it will provide the kind of compliance required for a specific task. Even so, there is no substitute for an actual test with the real tool on the manipulator.

-3-

Sensors

Sensors are completely unnecessary in NC tools but have a definite role to play in robotics. In order to program an NC machine, the location of every object involved in the machining process must be known. In NC turning, for example, the only objects that matter are the cutting tool and the blank. Both of these are held rigidly in position by the chuck and the tool post, so there is no difficulty in deciding where the two parts will be during cutting.

In a robotic work station, knowing where all the equipment will be is more difficult. First, there will simply be more pieces of equipment to keep track of. Second, at the time a task program for the station is being planned, that station may be working on a completely different task. The equipment being used in the ongoing task may have to be moved around in order to perform the new task, so knowing precisely where things will finally be located may be difficult. A third reason is that some equipment will inevitably fail during production and will have to be removed for maintenance or replacement.

Management must decide whether to expend once the effort necessary to develop software adequate to deal with inexactly positioned tooling or to design all the tooling so that it cannot be mispositioned. The alternative to extremely precise tooling is the use of sensors for determining the actual positions of things to within that accuracy. With the exception of television cameras, most sensors are extremely inexpensive in comparison to the cost of a manipulator (about $70,000 today). Even television cameras only cost about $1,000. The only argument against the use of sensors is whether or not the cost of software and computers needed to make use of sensors is prohibitive. Sixteen-bit microcomputers with 32K-word address spaces cost but a few thousand dollars today, and their price is dropping while the price of mechanical and electromechanical equipment is rising. Demonstrations by various industrial laboratories and pure research laboratories have shown that such computers ought to be perfectly adequate to handle all but the most demanding manipulator-control tasks in aerospace manufacturing when properly programmed. These computers could do many of the more demanding tasks with the aid of inexpensive computational hardware and properly designed tooling.

Sensors that will prove useful in automated aerospace manufacturing can be classified into four distinct categories: proximity, range, tactile, and visual sensors. The next subsection discusses some general principles to keep in mind when using sensors, and the following subsections discuss the four categories of sensors and give examples of specific sensors from each.

3.1 GENERAL CONSIDERATIONS FOR USE

The use of sensors in an automated aerospace manufacturing station affects the way in which programs to control that station must be written. Signal processing techniques can improve the performance of certain kinds of sensors regardless of the principles upon which they operate. These subjects are addressed in the following subsections.

3.1.1 Programming and Sensors

The task program for a work station can use sensors available at that station in order to obtain information on which to base decisions on which alternate processing steps to carry out. During normal production, the bulk of sensor readings that will be made will probably be for the purpose of verifying the correct completion of an individual processing step, such as drilling a hole or setting a rivet.

The task program can only obtain this information at run time after attempting the processing step. The program can then take some corrective (or at least protective) action if something went wrong. Present practice (and that mainly in research labs) is to develop algorithms for this sort of in-process testing in an ad hoc way and by writing fairly explicit task programs. The program-development process usually involves much imagination and tedious experimentation in order to determine whether the tests being made will detect enough of the processing errors that actually occur and whether the canned responses to those errors are adequate. In the future, as the aerospace industry begins to settle on standards of practice for robotic manufacturing, the problem of generating reliable task programs will become simpler because the number of choices that have to be made manually in deciding how to assemble a wing section, for example, will become fewer.

3.1.2 Teaching and Sensors

Aside from obtaining decision-making information, the other major use of sensors in an aerospace work station will be to supply, either indirectly as a result of intermediate computations or directly, the values of any deferred data items in the task program. The most common kind of deferred data in a task program will probably be position information. After that, visual information would probably be the next most frequently trained kind of information. The actual amount of visually input information could be quite large, however. Force and torque levels may not have to be trained very frequently at all. These levels will more likely be worked out during planning from known workpiece and tool weights and standards of practice and then supplied as predefined data values with the test of the task program.

Position information is very easy to train because a manipulator is in fact a large coordinate measuring machine. A special end effector shaped like a pointer will make it easier for the trainer to designate locations in the workspace whose X-Y-Z positions should be recorded. The work-station computer can easily compute the X-Y-Z values from the shape and size of the end effector, the arm's joint position, and the arm's geometry. For maximum accuracy, contact forces acting on the pointer cannot be allowed to deflect it. Even if the pointer end effector is very stiff and rigidly attached to the wrist socket, that force may still introduce measurement errors because of the small but finite amount of compliance in the manipulator. If the manipulator (and built-in software) permits access to the actual current position of each manipulator joint, then contact forces can introduce no measurement error. If, however, the only joint position data obtainable are the intended joint positions (i.e., the position set points of the encoders), then the contact force can cause a measurement error. The error will arise from the steady-state position errors in the individual joint servos; one or more of the joints will be a small distance away from their commanded positions essentially because the pointer end effector is blocking the path along which the wrist is trying to move.

3.1.3 Noise Immunity

A noncontact sensor is usually susceptible to interference from equipment that emits the energy to which the sensor responds -- light, sound, electromagnetic radiation, etc. This poses the problem of separating a signal from noise. Three general principles that are useful in increasing the sensitivity of such a sensor and reducing its susceptibility to noise and interference are filtering, modulating, and averaging. These principles can be used with sensors that respond to energy fields such as light, sound, magnetic, electrostatic, and radio-frequency emissions.

The principle behind filtering is that of screening out most of the noise energy on the basis of some property such as its frequency and concentrating as much as possible of the signal energy in the pass band of that filter.

The principle behind modulation is also that of filtering, but of filtering information that is carried by or encoded into the sensed energy field (which may itself be subject to filtering as described above). Modulation varies some aspect of the field (e.g., its strength, frequency, or spatial distribution) in a way that is unlikely to occur in the noise.

The principle behind averaging is to screen out noise on the basis of its randomness over a period of time. The signal should have some nonrandom properties that in some sense will not average out to a zero value.

For example, suppose the sensor is a photocell that is to respond only to light from a particular light-emitting diode (LED). At the photocell, filter out any light of a different color from that emitted by the LED. Modulate the light by turning the LED on and off 1000 times per second (this rate should be harmonically unrelated to the frequencies at which extraneous light might impinge on the photocell, such as the 60 Hz power-line frequency used in the United States). To detect light from the LED in the presence of other more intense light, sample the brightness of the received light with the photocell during both its on and off periods. The difference between the brightness sensed when the LED is on and the brightness sensed when it is off then indicates how much of the light being received by the photocell actually comes from LED. In order to reduce sensitivity to stray light further, average those differences over 10 or 100 successive samples (Reference 25).

3.2 PROXIMITY

A proximity sensor is a device that senses when one object (usually an end effector) is close to another object. Close can be anywhere from a few inches to a millimeter, depending upon the sensor used. Most of these devices indicate only the presence or absence of an object within their sensing region, but some can give some information about the distance between the object and the sensor as well. The following paragraphs describe several kinds of proximity sensors that could be useful in an aerospace manufacturing work station.

Optical-proximity sensors that are readily available on the market operate on either visible or invisible (almost always infrared) light. Most measure the amount of light reflected from an object. A factor in reliability is the type of light source that they use. The infrared-reflectance sensor with an incandescent light source is one of the most common. This sensor is widely available in a variety of convenient rugged packages and is not overly expensive.

Most optical sensors require a source of light. Incandescent filaments operated at reduced voltages can have multiyear lifetimes but are susceptible to damage from vibration. Light-emitting diodes have the reliability that is characteristic of other solid-state devices; they are insensitive to shock and vibration and are probably to be preferred over incandescent lights. Other light sources, such as electroluminescence or radiation-induced fluorescence, are not used much. Laser diodes can emit milliwatts of coherent light, but at present they are extremely expensive and their reliability is not as well established as that of other light sources.

Eddy-current proximity detectors produce an alternating magnetic field in a small volume of space at the tip of a probe. This field induces eddy currents in any conductive body that enters the sensitive volume. The eddy currents produce their own magnetic field that opposes the field emitted by the sensor. Coils or solid-state magnetic field sensors in the probe detect any change in the flux density at the probe tip and signal the presence of an object. The sensitive volume is usually quite small so that eddy-current proximity detectors are appropriate for detecting the presence of objects only when they approach the probe tip to within about a millimeter.

Magnetic-field sensors are very useful proximity detectors. These sensors may be made from a reed switch and a permanent magnet (in the object to be detected). Alternatively, the magnet may be part of the sensor, and the presence of the object can complete a magnetic circuit that operates the reed switch. Other forms of flux sensor, such as Hall-effect devices and magnetoresistive elements usually integrated with a solid-state amplifier for increased signal output, may also be used. The combination of a reed switch and a permanent magnet is particularly attractive because neither device must be supplied with power for operation.

Proximity detectors that operate on the basis of electrostatic effects can be built. The difficulty with these detectors is that they are quite sensitive to stray fields radiated by the electrical equipment and to fields from static charges induced by friction or by spraying operations. The signal conditioning and processing techniques described above might improve the performance of such sensors.

The familiar touch-sensitive button used in modern elevators can also be adapted for use as a proximity detector. In some of these devices, the capacitance between the person's body and his surroundings changes the resonant frequency of a tuned circuit. Usually, these devices only react to contact with a large conductive object, such as a person. However, by attaching a conductive plate or rod to the contact point, the device can respond to objects at a distance by virtue of their self-capacity.

Fluidic proximity detectors usually operate on the back pressure created when the presence of an object blocks an exit orifice. These devices can provide surprisingly precise indications of extremely small clearances between the probe and the object. These devices are in fact routinely used as sensors in automatic noncontact gaging and inspection equipment.

A novel acoustic proximity detector was recently developed, and it consists of a cylindrical open-ended reasonator cavity. An acoustic emitter at the closed end sets up standing waves in the cavity. The presence of an object closes off the open end of the resonator and changes the distribution of standing waves within the cavity. A microphone placed in the wall of the cavity detects the change in sound pressure as the standing-wave pattern moves. This device is also capable of precise measurement of the distance to the object (Reference 26).

3.3 RANGE

Although several have been built and demonstrated by various research laboratories, no commercial range finders of any use for aerospace processing are currently available. The term range sensor usually means a device that can provide precise measurement of the distance from the sensor to an object. Some of the capabilities that would be needed for aerospace manufacturing can be specified.

Ideally, the range sensor should require only a single line of sight so that it can look along the optical axis of a camera lens in order to produce range information that is in register with the image information from the camera (References 27,28). Having such information is a great aid in locating and identifying objects and is in fact the subject of research at the present time.

For aerospace industrial applications, the device should be able to measure distances from about one to about ten feet (0.3 to 3 meters), with a resolution of at least one part in 300 and preferably one part in 1000.

Such a device would be useful for locating objects within the work-station area and for controlling a manipulator. If, in addition, the device required only a small fraction of a second to measure a distance, it would also be useful for tracking moving objects and in line-following applications.

Only two kinds of commercially available devices can measure range at all. These are tellurometers and laser interferometric gages. The former are inappropriate for industrial applications because they only measure distances on the order of miles and only to an accuracy of about one foot. This device is typically used for survey measurements. Interferometric range measurement equipment can measure distances over the range of interest and with much better precision than is required in aerospace manufacturing, but it is extremely sensitive to environmental conditions such as humidity and temperature. Furthermore, those devices are usually not rugged enough to withstand rough handling or vibration; they are extremely expensive; and they require skilled operators.

A television camera can also be used to obtain range information by means of either stadimetric or triangulation methods (Subsection 3.5.2).

The Polaroid Corporation now markets a camera that is equipped with an acoustic range finder. This camera operates on a sonar principle. One problem with acoustic ranging devices is that their transverse spatial resolution is not very good because of the difficulty of producing a narrow beam of sound without elaborate equipment.

3.4 TACTILE

Tactile sensors respond to contact forces that arise between themselves and solid objects. Unlike proximity detectors, tactile sensors do not indicate the presence of an object until it actually touches the sensor. A useful combination of sensors in an end effector is a proximity sensor that works in conjunction with a touch sensor. The proximity detector can detect the presence of an object still some distance away so that the work-station controller can safely move the end effector quickly towards that object even if its position is not precisely known. The signal from the proximity detector would give the work-station controller the warning it would need in order to slow down and avoid a collision. The controller could monitor the touch sensor while moving the end effector slowly towards the target.

Tactile sensors can be classified into touch sensors and stress sensors. Touch sensors produce a binary output signal, depending upon whether or not they are in contact with something. Stress sensors produce signals that indicate the magnitude of the contact forces. Individual stress sensors usually respond only to force in one direction on them. However, combinations of two or more can report forces as well as torques in two or three directions.

The simplest kind of touch sensor requires no specific sensor device at all if the objects that they are going to touch are electrically conductive. Just apply a small potential difference between them, and when it goes to zero, contact has occurred.

Microswitches are probably the least expensive and most commonly used form of touch sensor. Microswitches should be mounted so that they are protected against accidental collisions with objects in the workspace. These devices can be equipped with feelers to protect them against excessive force and to extend the region in which they can sense contact.

Inexpensive tactile sensor arrays may be constructed from commercially available rubber sheets that have been doped with minute silver granules to give them electrical properties. The resistance across the sheet is normally quite high, as rubber is an insulator. When an object touches the sheet and compresses it, the resistance across such a sheet decreases abruptly. At a certain level of compression, sufficient silver granules to form conducting pathways from one side of the rubber sheet to the other are forced into contact with one another. Thus, electrical connections are formed through the sheet at each place where an object touches it (Reference 29).

Strain gages are often used to make force sensors, torque sensors, and sensors that can measure both kinds of stress simultaneously (References 30,31). The sensors are usually constructed by attaching individual strain gages to the roots of cantilever beams milled into solid blocks of aluminum. The orientations of the beams and the connections between them may be designed to resolve partially the applied force and torque mechanically into its six components with respect to a set of Cartesian axes fixed in the block. Alternatively, the beams may be positioned according to other criteria such as strength or convenience of manufacture. The various stress components may then be resolved by appropriate software (a process called diagonalization).

Shimano at the Stanford Artificial Intelligence Laboratory has demonstrated such a software technique (Reference 32). Shimano, using eight strain gages, formed an eight-element vector from the signals they produced and multiplied that vector by a six-by-eight matrix of sensitivity coefficients. He also demonstrated an elegant method by which the controlling computer could work out the values for those coefficients rapidly and without using any special mechanical or electrical measuring equipment. In this procedure, the sensor is mounted between a manipulator and an end effector as a wrist. The procedure uses the end effector's known weight in a fully automatic calibration procedure for the six-degree-of-freedom stress-sensing device capable of being carried out by the work-station controller without human aid. Transforming a set of forces and torques from one reference frame to another with software or appropriate analog computer hardware is a simple matter called remote moment sensing.

3.5 VISUAL*

Automatic computer vision will be an essential part of flexible automated manufacturing systems of the future; existing robot technology is clearly in need of sensory feedback to extend its limited capabilities. Special-purpose vision systems are already appearing in increasing numbers on factory floors. But, considering the premium that batch fabrication places on a plant's ability to respond quickly to managerial decisions and the vagaries of delivery schedules while processing a wide and ever-changing variety of parts, only general-purpose vision systems will provide both the requisite generality and processing power in the long run.

Visual feedback can minimize the need for jigs and fixtures and ease tolerances on parts. Visual feedback controlling a manipulator in real time can allow it to work on a moving line without requiring precise control of the line. The same vision system used for these purposes can also provide 100 percent process inspection capability for little or no additional investment.

Vision systems can be used for two different purposes: to recognize objects and to measure specific characteristics of the objects (Reference 33). The following subsections discuss methods for recognizing objects such as workpieces and three different types of measurements that should prove useful in manipulator control for aerospace applications: depth, surface orientation, and object position.

3.5.1 Recognition

Although the question of how to program a camera-equipped computer to recognize an arbitrary object placed in an arbitrary scene is still a meaty research topic, this is not a problem that robotic aerospace manufacturing has to solve. In a factory environment it is entirely practical to control such factors as the illumination, the background, the viewpoint, and even to some extent the position of the workpiece in order to simplify the image as much as possible and to emphasize the features that are most significant. In order to identify a part from a picture taken through a television camera, the part must merely be distinguished reliably from any other parts that might appear before that camera at that time.

The usual approach in distinguishing between several different classes of objects is to determine some specific characteristics of the given object, and then compare these to the corresponding characteristics of prototype objects, each representing the typical part in a given class. Luckily, objects commonly encountered in a manufacturing situation tend to have distinctly different shapes and sizes and objects of a given class, perhaps a blank for a specific interchangeable/replaceable (I/R) panel, are almost always rigid and very similar in their dimensions. Recognizing objects in the aerospace manufacturing environment is considerably simpler than other classic (and very difficult) pattern-recognition problems such as finding the cancerous cells in a Pap smear, reading hand-painted letters, and understanding spoken English.

One practical approach to industrial part recognition is to measure a set of features that tend to remain constant for solid objects of almost identical shape (a good description of most industrial workpieces). Some appropriate features are the area of the object's image, its perimeter, major and minor axes, second moments of

---

* The material in this section is adapted from two papers by G. Agin and G. Gleason of SRI International (References 34,35).

area, and so on. Such features are especially good for distinguishing between parts if the values of those features are invariant to the parts' rotations around an axis parallel to the camera's direction of view. Such features then allow the part to be recognized regardless of its orientation. In industrial applications, control of the part orientation usually goes hand-in-hand with knowledge of that part's identity. Recognition of disoriented parts is important because they are more likely to be unidentified.

Other features that may be measured depend upon the part's position and orientation, such as the minimal bounding rectangle around the image or the direction of an axis of symmetry. These features are useful for determining the orientation of the part. The part's location is usually obtained by simply finding the center of gravity of the image of that part. Occasionally, lighting conditions or something about the material or finish of the part itself will make it difficult to obtain repeatable images of that part. In that case, other more sophisticated tests on the features may be needed in order to measure its location.

The camera can locate any object in its field of view to any accuracy within the limits of its resolution. The physical design of the camera places a restriction on the resolution across the field of view, and resolutions of about 100 picture elements (pixels) across the field are common today. Because modern solid-state cameras are produced by integrated-circuit masking techniques, linearity across the field is not a problem.

What determines the spatial resolution of the camera at the workpiece is really the lens that one places in front of the camera and the distance from the camera to the workpiece. A television camera can locate an object to within a small fraction of a mil if it looks at that object through a microscope. If the position of the camera is also known, the camera will deliver the absolute position of that object with the same high accuracy. The problem is that with a strong lens, the field of view shrinks. A proposal frequently made is to equip a television camera with a zoom lens that can provide any magnification from wide-angle to close-up and to direct its gaze with a two-axis galvanometer mirror. Such a camera (foveal camera) could take a wide-angle view of the work area in order to locate an object roughly, then zoom the area for successively closer looks until it had located the object with the required accuracy. Similar results can be obtained with a camera on the end effector and with automatic control of the lens focus.

3.5.2 Depth Measurement

The function of computer vision in a visual servoing application is to determine the spatial relationships that exist between the camera, the end effector, and the workpiece. Two methods by which a single camera can obtain depth information from a scene are stadimetry and triangulation.

Stadimetry is the process of inferring the distance to an object on the basis of its apparent size in the image. Of course, this requires that the visual system must locate the object in the field of view and identify the object uniquely from among other similar objects in the scene.

Some of the practical difficulties in applying stadimetric methods are that the size of the image will vary with the focus of the lens and setting of the binary threshold. A way to get around this difficulty is to measure the distance between two stripes, spots, or holes on an object; defocusing and variation in threshold level will not affect the apparent locations of the centers of those marks very much.

Triangulation methods are based on measuring the angles and base line of a triangle whose apex is at the location of the object whose distance is to be determined. The sides of the triangle may be formed by lines of sight toward the object from different camera positions, either simultaneously by different cameras (stereogrammetry), or at different times by the same camera (motion parallax). Alternatively, one or both sides of the triangle may be formed by projected beams of light with a camera viewing the spot(s) made by the light beam(s) falling on the object.

The major problem with triangulation methods is occlusion; some object in the workspace may intersect one of the lines of sight or light beams and block the view of the camera or cast a shadow on the target object. A good example is the problem of determining the depth of a hole by triangulation methods. If the hole is very deep, the base of the triangle must be quite small so that the two sides of the triangle do not touch the sides of the hole. Therefore, obtaining any kind of accuracy in the depth measurement requires extremely accurate measurement of the angle that each side of the triangle makes with respect to the base line.

The Konica Corporation has recently placed on the market a 35mm single-lens reflex camera containing a triangulation range-finding system that enables the camera to focus itself automatically (Reference 36). The range finder consists of a special-purpose integrated circuit that correlates the brightness values in the two scenes viewed through a conventional split-image range finder. In effect, the circuit is able to tell when the image in the two halves of the split image are lined up. The integrated circuit promises to have many applications for range finding in industrial settings.

3.5.3 Surface Orientation Measurement

Simple patterns of light projected on the workpiece can give additional information about the location and orientation of an object, depending upon the pattern and the shape of the object (References 34,37,38). For example, two parallel vertical stripes can give information regarding the rotation of a plane about a horizontal axis in the scene. The possibilities are too numerous to mention, but alert robot programmers will quickly see applications for techniques like this one in specific aerospace tasks.

3.5.4 Position Measurement

A small, rugged, solid-state television camera may be placed in the manipulator's end effector, and its visual feedback may be used to guide the hand to a given target. This procedure is called visual servoing. Visual servoing could be applied to a large variety of tasks in material handling (moving parts from place to place), fitting (aligning parts with respect to one another), fastening, and machine tool loading.

Little prior effort has been made in visual servoing of a manipulator. A few experiments have been performed in the "blocks world" in which the only objects are smooth, clean, regular solids such as cubes and pyramids (References 30,34,39-42). In these experiments, a fixed television camera observed a robot at work. Attempts to place one block on top of another or to insert a peg into an oversized hole were made by carefully observing an area of the scene where the mating would take place. The real-time control aspect of visual servoing has been absent in these experiments where the basic method was to alternate repeatedly between taking pictures and moving the manipulator.

For industrial application, the approach has generally been to move the television camera with respect to the workpiece. In one experiment of this nature carried out in a Japanese laboratory, the camera was rotated and translated until the perceived image was properly aligned and centered. A more recent publication, also by a Japanese research team, describes visual servoing with both fixed and manipulator-mounted cameras (Reference 43).

For real-time control using visual feedback, a key point is to make use of binary images only because they can be processed more quickly and reliably than grey-scale images. The constraint of binary image processing forces special consideration for lighting and contrast in the image, but the reward for this is fast operation. In some applications, using projected light patterns will be practical to obtain information about range or depth. The real-time nature of the servoing problem requires consideration of the dynamics of mechanical components and leads to questions of stability and speed of response.

Two distinct modes of visual servoing are the point mode and the line-following mode. In the point mode, servoing is used to bring an end effector to some specific location - for instance, to insert a rivet into a hole. If the target is in motion, the servo system should track it so that the relative velocity of the camera and tool with respect to the workpiece is zero. In the line-following mode, the objective is to follow a path at some specific nonzero velocity -- for example, in tasks that require sealing, gluing, or seam following.

When the camera moves with respect to the target in the line-following mode, additional geometric information about the position and orientation of the object may be obtained from that motion. Successive images of a groove can (with knowledge of how the camera moves) give the orientation of the groove where a single image cannot.

Once visual servoing attains the desired relationship between the camera and the workpiece, the end effector can move a fixed distance and place itself in the same relationship to that workpiece. In point-mode servoing, this procedure involves a separate motion. In the line-following mode, however, the end effector can simply follow the camera.

If the visual target is stationary in the camera's field of view, then performance of a task will require only positioning of the end effector in the field of view. In following a path at a fixed velocity, however, the work-station controller should have the ability to deal with moving coordinate systems. With this ability, only maintaining a fixed velocity (again relative to the camera's coordinate frame) is necessary; thereafter, the controller can command position changes relative to that moving frame.

The servo systems that operate the joints of commercial robots are rather complicated mechanisms. Their responses are generally nonlinear, nonisotropic, and load-dependent. These robots can all go to any commanded position in a reasonable time, but some of their joints may arrive at their final positions faster than others and may not arrive in the same order each time. The motion of some joints can be dramatically affected by simultaneous motions of the other joints due to inertial coupling between the varous links of the manipulator. Nevertheless, with sufficient care in the design of the visual servo loop, satisfactory results can be obtained.

The simplest way to servo a manipulator visually is to take a single picture, estimate the position error, calculate a new position that will reduce the error, and command the manipulator to go to that position. Wait a sufficient time for the manipulator to complete that motion, then repeat the process. When the target is stationary and the speed of response is not critical, this approach can give quite adequate results.

When a faster servo response is desired, or when the target may be moving in an unpredictable way, taking pictures as often as possible becomes desirable. If, for each picture, the work-station controller were to calculate an incremental movement that would precisely cancel out the position error observed through the camera, the various unavoidable delays in the manipulator's response would quickly cause a highly unstable response. A way to defeat this instability is to command smaller moves. The correction applied to the end-effector position (as seen by the camera) might be computed as some factor (beta) times observed position error. In that case, beta should be always less than or equal to 1.0 in order to avoid overshooting the target and guarantee oscillatory end-effector motion. If beta is very small, however, response will be too sluggish (underdamped) and throughput will suffer.

If the target is moving unpredictably, the position error observed by the camera may also be used to produce an estimate of its instantaneous velocity. For example, each time the target appears to the right of the center of the camera image, the work-station controller might increase its estimate of how fast the target was moving to the right. To keep up with the target, the work-station controller would then increase the velocity at which it was moving the camera/tool end effector to the right.

Again, one convenient and simple algorithm for estimating the velocity of a target might be to increase the estimate by the product of the observed position error and a factor (gamma). Gamma is thus the change in the estimated velocity per unit of position error during the time between successive pictures. This servo algorithm, depending on the values of gamma and the camera's frame rate, can also display underdamped, critically damped, or oscillatory behavior. If one should wish to try out the algorithm, one should take pictures at the highest rate that can be processed and adjust the value of gamma experimentally to obtain a critically damped response. Gamma has the dimensions of

$$\frac{\text{(distance/time)}}{\text{distance}} \quad \text{or} \quad \frac{1}{\text{time}},$$

and its value determines how fast a target can move and the system still acquire the target (begin tracking it and continue to do so). The system should be able to track successfully any object that enters its field of view at a constant velocity that will make it take longer than 1/gamma to cross the camera's field of view. The system will need to catch the target in two successive pictures in order to determine the direction that the target is moving. The camera will have to take at least one picture every 1/gamma seconds. If the camera takes pictures at a faster rate, the acceleration of the end effector as it begins to move with the target will be smoother and the system will be more apt to acquire the target successfully. A quantitative prediction of the tracking behavior for various combinations of gamma and the picture-taking rate would require a sampled-data analysis, and the effects of the manipulator's dynamics probably would have a significant effect on stability.

Another factor to consider in visual servoing is the finite time that will be required to process any image. An unavoidable delay in the feedback path of the servo is another potential source of instability. Various techniques may be used to mitigate the effects of this delay, but the faster an image can be processed, the better the tracking performance will be.

Another important factor to consider is that it is vital to know exactly where the camera is whenever it takes a picture. Whenever the work-station controller is trying to track a moving object with a camera held by the manipulator, the individual joints of the manipulator will generally not be at their last-commanded position (i.e., there will be some small position errors in one or more of the joint servos). Therefore, accurate interpretation of data from a manipulator-mounted camera depends upon obtaining on demand from the manipulator (or from its control system) accurate information about the actual position of every joint. If the manipulator can supply this information, the correct procedure for locating an object with a moving camera is to take a picture and, as simultaneously as possible, ask for the manipulator's position. The position of the camera when it took the picture could then be computed, and the image can then be analyzed in relative leisure.

Finally, if the camera and manipulator are operated by different computers, any delays introduced by the communication channel between the two machines will encourage instability.

-4-

Tooling

Tooling can be divided into four classifications - fixed, movable, passive, and active. Each of these classifications poses characteristic control problems for the work-station control computer. Sensors and powered tools are further classified on the basis of how they interact with workpieces. The characteristics of specific end effectors are discussed in Subsection 4.2.2.

4.1 CLASSIFICATION OF TOOLING

In the following discussions, tooling is classified as being either fixed or movable and either passive or active. Fixed tools always sit in one place while movable tools can be carried around by a manipulator. Passive tools contain no actuators or sensors and exchange no signals with the work-station control computer while active tools do. This breakdown gives the four classes of tooling shown in Figure 11. Each class poses significantly different control problems.

Fixed passive tooling includes all objects capable only of supporting another movable piece of tooling (including a workpiece). This category includes jigs, work tables, and tool racks.

Fixed active tooling includes all equipment that requires control signals, produces information, and is not moved from place to place by the manipulator. This category includes conveyors, conventional NC equipment, part feeders, vises, clamps, furnace doors, part orienters, and glue dispensers. Any sensors such as photocells, proximity detectors, scales, force-sensing tables, and cameras that are mounted permanently in one place also fall into this class.

Movable passive tooling includes all unpowered objects that the manipulator can move from place to place. This classification includes tote boxes, templates, fasteners (e.g., rivets and jigging components such as Cleco clips), and the workpieces themselves.

Movable active tooling includes all objects that the manipulator can pick up and move from place to place and that either require control signals or produce information. Tooling in this class may perform its function while the manipulator is carrying it. Some tooling in this class includes the robot's gripper and any sensors or power tools that the robot carries (possibly in a gripper), such as drills, spot welders, spray guns, force-sensing wrists, cameras, optical character readers, or range finders.

Two other factors of active tooling, both fixed and movable, determine the difficulty that the work-station controller will have in operating the tooling. These two factors are whether the tool makes contact with the workpiece and the dimensionality of the region over which the tool interacts with the workpiece. This leads to the eight-way classification scheme shown in Figure 11, under movable active tooling.

Tooling that touches the workpiece includes grippers of all kinds as well as most kinds of tools that remove material from the workpiece or change the shape of the workpiece. Tools which do not contact the workpiece include most tools that deposit material. A sensor can be used as a special kind of tool that has no effect on the workpiece. Sensors are often included as components of multiple-purpose tools, and the sensors themselves may be either contacting or noncontacting.

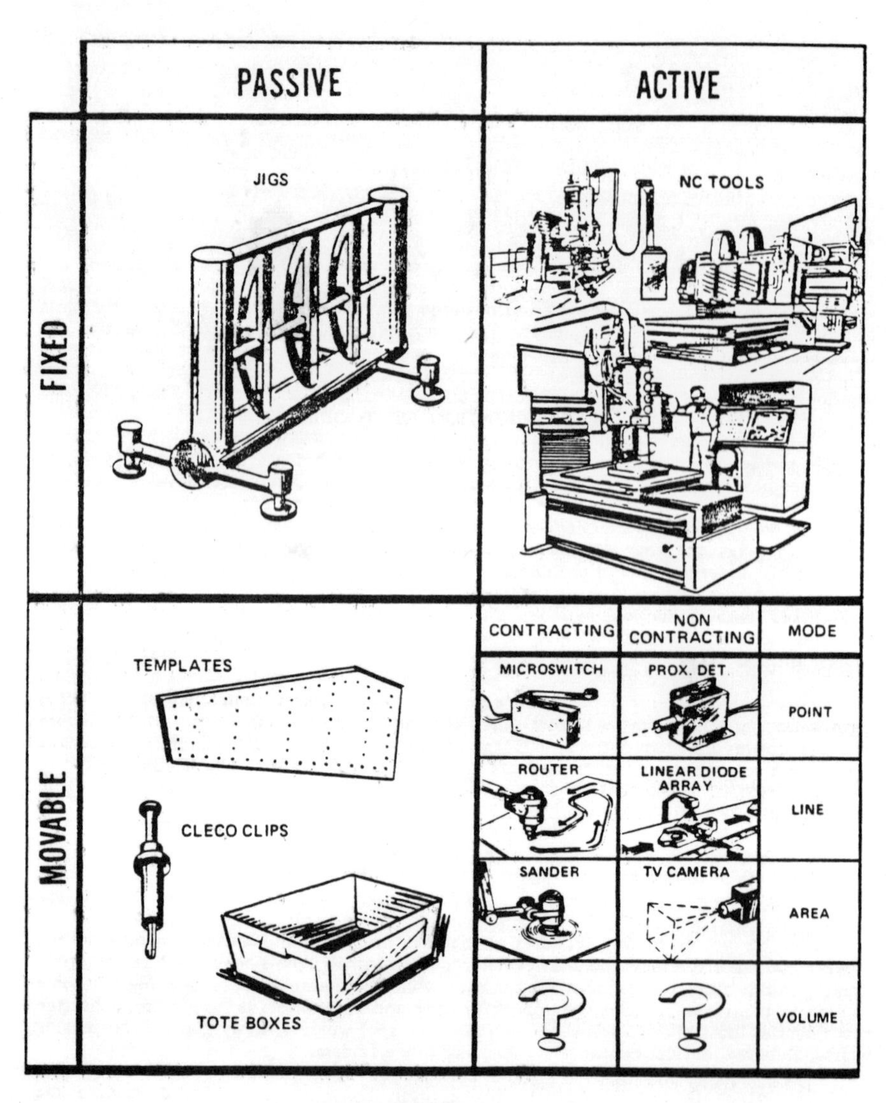

Figure 11
CLASSES OF TOOLING

An active tool also tends to interact with a workpiece in primarily one of four ways: at a point, along a line, over an area, or throughout a volume. Of course, the interaction regions are not really ideal mathematical points, lines, or planes. For example, a drill can be classified as a point-processing end effector even though it clearly removes a volume of material from the workpiece. Similarly, an edge router can be called a line-following end effector, and a spray gun can be called an area-covering end effector. In the aerospace industry, the removal of volumes of metal is accomplished almost exclusively by traditional NC machines. Since there are standard methods such as APT programming for controlling NC tools, volume-processing tooling will not be discussed further.

4.2 END EFFECTORS

Most of the significant control problems in a work station will have to do with the operation of movable active tooling that performs its function while the manipulator carries it. This type of tool is called an end effector. An end effector is any device attached to the end of the manipualtor to perform one or more functions such as sensing, gripping, or a manufacturing process such as drilling, routing, or spraying.

4.2.1 General Considerations

End effectors usually fall into the class of movable active tooling. Only a few examples of passive end effectors (such as ladles, which have been used for pouring molten metal into molds) exist. In the following discussion, the term end effector is used to mean either a gripper, a tool held by a gripper, or a tool mounted on a manipulator's wrist. The definition depends upon the context of the discussion.

Much work goes into the design of end effectors. The effectors must be rugged enough to withstand occasional accidental collisions. If the end effecors are too heavy, they will slow down the manipulator and reduce the load that the manipulator can carry. If they are too big, they may not be able to reach between obstructions to get to the workpiece. In order to reduce the amount of time wasted in changing tools during production, an end effector may be designed to perform several different functions. Except for simple grippers, end effectors are usually custom designs worked out by the manipulator's owners. Most manipulator manufacturers will advise their customers on end-effector design.

4.2.2 Characteristics of Specific End Effectors

Several kinds of end effectors were identified as being particularly important in aerospace manufacturing on the basis of the responses to a questionnaire circulated by the McDonnell-Douglas Corporation. The characteristics of these grippers, welders, grinders, deburrers, painters, routers, and drills are discussed in the following subsections.

4.2.2.1 Grippers

Grippers are used for two distinctly different purposes: for performing pick-and-place operations and for holding tools that perform processes on workpieces. Although a tremendous variety of gripper designs are in use, most designs grasp objects either with suction cups, magnets, or articulated mechanisms. Some grippers contain their own actuators that allow them to move or rotate objects without requiring the manipulator to move.

In reading the following sections, please keep in mind the fact that any gripper can also be used as a vise and a clamping mechanism. Included in the work station tooling a "live" end-effector socket mounted in a fixed location on a workbench or jig just for such uses may be worthwhile.

Grippers are inherently contacting point-processing tools. This has important implications for how the work-station controller should use them. Normally, a gripper is used to pick objects up and put them down. A gripper usually holds an object so that the object can neither translate nor rotate relative to the gripper (i.e., the gripper constrains all six of the object's degrees of freedom). Some grippers are designed with built-in compliances, but many are simply attached rigidly to the manipulator's wrist. In order for the work-station controller to grasp and release objects reliably with a rigid gripper and without exerting excessive forces on the objects, all constraints that will be imposed on the position and orientation of a workpiece during handling must be taken into account.

Grippers generally use one of four methods for holding an object: friction, physical constraints, attraction, or support. Friction and constraining grippers are usually linkages (jointed mechanisms) operated by one or more actuators, some of which may be servoed. These grippers may also be constructed with inflatable bladders in various configurations to grip parts of particular shapes.

Friction grippers exert pressure on a workpiece, either by expanding within it or by closing on it from outside. The workpiece can be pulled away from such a gripper with sufficient force; this feature can actually be a safety factor in some applications. Friction grippers generally rely on soft materials at the point of contact wth an object in order to give sufficient force of friction for a secure grasp. Material that will remain soft under repeated impact loads and that is oil-resistant (if the manipulator is hydraulically-powered) should be chosen. Some newly developed materials that have extremely high coefficients or friction may be useful if they prove durable enough. Any such soft materials are subject to wear and should be replaced whenever they become damaged so that pieces of them will not fall off into the aircraft.

Physically constraining grippers may or may not exert pressure on a workpiece. Instead, these grippers grasp the workpiece by placing solid material around it in order to prevent it from moving. Most of these grippers hold a workpiece rigidly, but in one popular design suitable for light-duty use, prehensile elastomeric fingers curl gently around the workpiece when high-pressure air is pumped into them (active end mechanism). The Japanese have pioneered in the design of tentacle-like mechanical linkages that wrap themselves around the workpiece and conform to its shape (References 44,45). Their actuating mechanisms are surprisingly simple and reliable. An unusual kind of gripper (or vise) for objects of unpredictable shape can be made with a granular material, such as sand or magnetic particles, in a loose bag. Draping the bag over a workpiece and applying a vacuum or magnetic field gives the powder sufficient rigidity to support the workpiece when the bag is lifted. A magnetic fluid could also be used. Fluidized beds of sand or ball bearings can be used as vises or clamps with vacuum or magnetism as an aid in rigidifying the medium (Reference 46).

A tremendous variety of clever linkage designs have been used in grippers, but their overall action can be identified as being either parallel-jaw, two-fingered, three-fingered, or multi-fingered. Parallel-jaw grippers contact the workpiece over relatively large areas by bringing two flat surfaces together on opposite sides of it. Finger grippers usually make contact at relatively small regions.

Probably the most versatile kind of gripper is a Skinner hand, named inventor, Frank Skinner (References 47,48). Skinner suggested a three-fingered design in which each finger is capable of prehension, and a joint at the base of each allows it to twist about its long axis as shown in Figure 12. This kind of hand could be used for friction, physical constraint, and support modes of gripping, such as the power grip, the two-fingered pinch, and the suitcase carry. Unfortunately, such hands would probably be complex mechanisms and consequently be expensive to construct. A commercial line of standardized, industrial-quality Skinner hands would be useful to the robotic community. Unfortunately, the market for such grippers may be too small to justify their development costs.

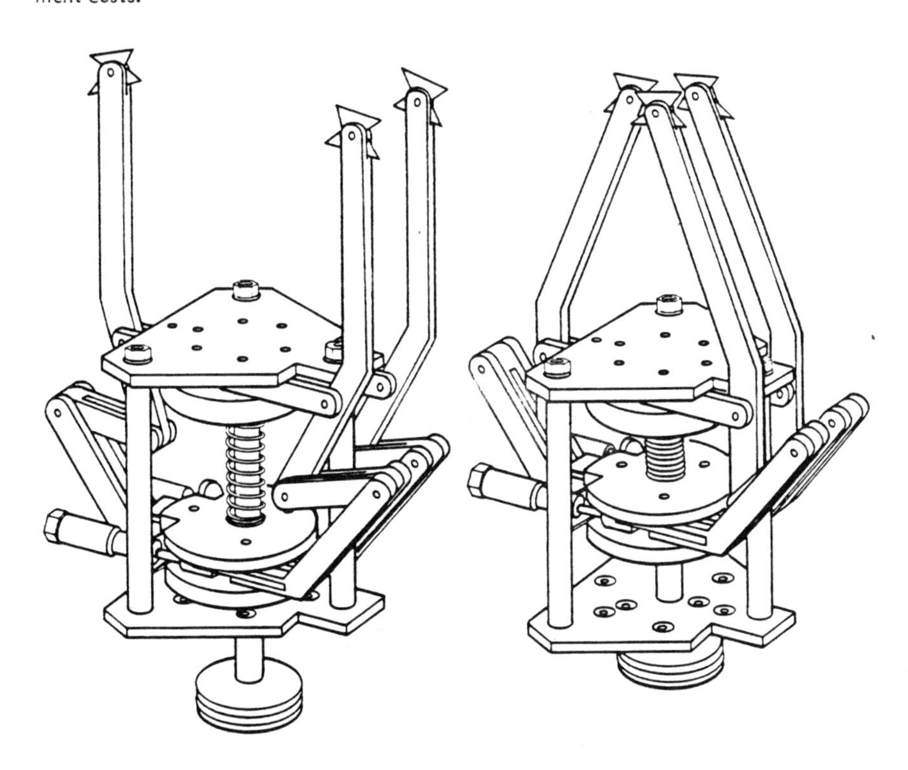

Figure 12
SKINNER HAND

2) Attraction grippers use either magnetic force or suction to hold an object. Suction cups will probably be more useful to aerospace manufacturers since most of the material they handle is nonmagnetic and in sheet form. Adhesion has not been used much in grippers to date but could well be. Adhesive coatings on pallets have already been used in factory environments in order to help parts retain their orientations during transport. Aerospace manufacturing applications would probably require gripper adhesives with very low transfer coefficients. Arrays of magnets or suction cups on compliant mountings are useful in grasping irregularly shaped workpieces. The standard practice to reduce the air-flow rate required to operate a large array of suction cups is to place a valve in each cup. The valve connects that cup to the vacuum line when the cup contacts an object. Placing an orifice between each cup and the line is a less effective but cheaper solution to the same problem.

3) The most widely used form of support gripper is a hook, and this gripper is usually found only on crane-type manipulators. A hook can be a useful accessory on a manipulator when the hook is being operated under remote control during the equipment setup for a batch production run, for example. There is a tendency to avoid gripper designs that only support a workpiece from below because the workpiece can easily fall out of or off of such a gripper when the manipulator moves quickly. The standard practice for moving an object is to constrain all six degrees of freedom of the object being gripped.

4.2.2.2 Welders

Welders may be either contacting or noncontacting tools; the work station always has to issue some sort of controls to operate them so they are also active tools. Some types of welders used in aerospace applications are spot welders, stud welders, stick, MIG or TIG welders, and plasma arcs.

A spot-weld gun is a point-processing tool; it grips the workpiece tightly between two jaws or horns, and all the manipulator has to do is position the gun correctly with respect to the workpiece, turn it on, wait for the gun to complete the cycle, and open its jaws. During welding, the jaws constrain the position in space of the weld point on the workpiece. The workpiece, therefore, can only rotate around the spot on which the jaws closed. This set of constraints can make the spot welder exert damaging levels of thrust on the workpiece if the manipulator should move relative to the workpiece during the welding cycle. Sometimes the horns will stick to the workpiece; this can be detected. The usual response to this undesirable condition is to twist the entire gun rapidly back and forth in order to break the unintentional weld between the metal and the horns. This, of course, unavoidably exerts a lot of force on the workpiece, and it may damage thin sheet metal parts.

A MIG, TIG, or stick welding gun exerts little or no force unless the electrode or make-up rod accidentally touches the work. Plasma-arc guns have no electrodes and exert very little force. All of these arc welders are usually used as line tools, so the work station controller must make their tips follow precise trajectories in space and time in order to lay down good weld beads. One complication is that the weld gun often has to move at a varying speed along the weld seam. The proper speed depends in a complex way on many factors such as the shape and thickness of the metal nearest the arc, the local radius of the path curvature, and the instantaneous size of the puddle. The workpiece may distort from heat, and sensing anything near the electric arc, whether with optical or other kinds of sensors, is quite difficult (Reference 49).

4.2.2.3 Grinders and Deburrers

Many grinding and deburring operations are low-precision processes that can be accomplished by pressing a rotating abrading tool against a workpiece and dragging it along a path over the workpiece's surface. The force on the tool is large and variable. Using a template to guide the tool may be practical, as in edge routing (Subsection 4.2.2.5). In this case, the control problems are those of template following (Subsection 5.2.4.3.2).

Grinders and deburrers can be considered line-following tools although they sometimes are used to smooth a surface. Precision grinding and deburring are usually done on conventional NC tools. Polishing and sanding are similar operations, usually on surfaces, but they are not very frequent operations in aerospace manufacturing.

Operating a grinder or deburrer requires that the work station keep a rotating tool face pressed against the workpiece and that the tool be able to comply in a direction normal to the surface or edge being followed as it passes over lumps of material to be removed. At the same time, the manipulator must resist side forces or torques that arise from friction between the rotating tool face and the work surface. If the manipulator does not resist, various kinds of chatter and vibration which can damage the workpiece may occur. Deburring tools usually remove all burrs in one pass. For reliable and efficient grinding, however, the work station may have to monitor the shape of the workpiece in order to ensure that all excess material is removed and that time is not wasted in grinding where there is no flash.

4.2.2.4 Painters

Spray painting is an important area for automation because it presents major health and safety hazards when it is done manually (Reference 50). In cold climates, spray painting also requires significant expenditures for heating the air in the painting booth. The air must be constantly replaced in order to reduce the solvent concentration enough so that the painters can breathe it. Techniques developed for spray painting may also be applicable to other processes in which a quantity must be applied in a controlled way to a large surface. Two such processes are the application of protective films to sheet metal and the application of heat to foundry molds for bakeout.

Paint is always applied with a spray gun in automated systems. In principle, there is no reason why a manipulator should not wield a roller or a brush if it can apply enough paint to be cost-effective. If possible, the work-station controller should present the various surfaces of the workpiece so that the manipulator can easily reach any point on them and so that the workpiece is stationary. Flat and cylindrical surfaces are easy to paint with a simple scanning motion of the manipulator. The doubly curved surfaces that occur frequently in aerospace manufacturing may require more complex spray-gun trajectories in order to obtain a uniform coating of paint because of overlap effects.

Spray painting in the general case presents several challenging control problems, such as

- o The coat of paint must completely cover the surface(s) to be painted.
- o The coat must be of a certain minimum thickness.
- o The coat must be of uniform thickness.
- o The amount of paint wasted must be minimized.

Little theoretical work has been done on methods for precomputing optimal spray-gun trajectories for computer-controlled spray-painting robots. The paint flux density in the plume varies approximately as the inverse square of the distance from the nozzle to the workpiece over a certain range of distances. However, the thickness of the coating depends upon the integral of the flux density across the plume in the direction of traverse. Therefore (over some range of distances), the traverse velocity should vary approximately inverse with the distance in order to maintain a constant coating thickness in one pass. A higher-level consideration is that for larger distances the width of the sprayed path will be larger; fewer passes will be required, and the passes should be spaced further apart.

Current practice in robotic spray painting is to record the spray gun motions while the equipment is under the control of a skilled human spray painter. The robot's controller then just replays these recorded motions verbatim. In this method, all the parts must be presented in the identical position and orientation as the part that was painted by the person. The painting of moving parts by this technique is possible as long as all the parts move in exactly the same way as the part that was painted during training. Claims that advanced commercial spray-painting systems can now paint moving objects after being trained on a stationary one have been made.

Because spray-paint guns are area tools, the work station must give due consideration to the effects of overlapping coverage. Furthermore, if the shape of the spray produced by the gun has a circular cross section, a manipulator with one less joint may possibly be used. This is because the orientation of the spray gun about the axis of its nozzle will then have no effect on the thickness or distribution of the coating of paint laid down. Control of the spray-gun orientation about that axis is then unnessary, and in principle, one less joint should be needed.

The general problem of spray painting an object that is presented in an arbitrary orientation, and possibly also moving, is one of the most difficult control problems in robotics for several reasons. The speed requirements usually will require the work-station controller to take into account the manipulator's dynamic limitations. Kinematically, this is an area coverage problem in which overlap affects matter -- the most difficult kind. Synchronous trajectory-following is required because any variation in the speed of the spray gun along the trajectory will affect the thickness of the coating of paint that is deposited. In order to guarantee adequate coverage, using machine vision to see where more paint is needed and to track the workpiece if it is moving may be necessary. Vision is difficult in a spray-painting environment. Industrial spray guns deliver paint at a tremendous rate and exert a considerable reaction force on the wrist. Since most spray-painting manipulators have comparatively lightweight flimsy links in order to obtain high accelerations, this force can lead to considerable dynamic control problems. The most difficult control problem of all in building a robot that can spray paint as well as a person may well prove to be the planning of the arm's motion. For example, the arm's joints must never exceed their individual ranges of motion, and the arm must never collide with the workpiece in trying to reach around behind it.

Because of all the difficulties mentioned above, it would be wise to structure the work situation in order to simplify the job of the work-station controller as much as possible (e.g., by only painting stationary parts in similar orientations).

4.2.2.5 Routers

A router is a line-following end effector that contacts the workpiece and exerts large forces on it. Routing is an important process in aerospace sheet metal fabrication because of the need to accurately shape field-replaceable skin surfaces of modern fighter aircraft. One-pass routing to within a few mils tolerance requires an extremely stiff tool mount -- much stiffer than commercial manipulators can supply. For this reason, routing is a good candidate for the template-guided method of line-following (Subsection 5.2.4.3.2).

4.2.2.6 Drills

A drill is a point-processing end effector that contacts the workpiece and exerts large forces on it. Drilling holes is one of the most frequent unit operations in aerospace manufacturing. The tolerance on hole placement is 30 mils or less, but this is often only the tolerance on the relatively small distance from the hole to the outside edge of a stringer (approximately .5-inch). Thus, in most cases, the accurate placement of a hole does not require extreme resolution in the position-measuring equipment. The work-station controller should be able to take advantage of this fact. Drilling is a good candidate for the template-guided method of point-processing. The drill must be oriented normal to the surface to be drilled. The work-station controller can do this by the template or by sensor-controlled orientation of the drill. Drilling is also a good candidate for smart tooling.

4.2.3 Robot/End-Effector Interfacing

One of the most useful and important decisions to be made in setting up a work station will be the design of the interface between the manipulator and the end effectors that it carries (References 51-53). The interface must support the end effector structurally, provide it with power, and convey information to and from it. The interface must be reliable and must be designed to permit quick connection and disconnection. The manipulator should not have to be positioned with extreme accuracy in order to make the connections. The interface equipment should be impervious to what-ever environmental insults are likely to occur in normal operations, such as moisture, oil, metal chips, and occasional collisions.

Power and information can be transmitted in several different ways, some of which may offer advantages over others in certain situations. For example, in fully automatic spray painting, operating the manipulator in an atmosphere whose solvent concentration is above the explosive level may be cost-effective. In that situation, selection of nonelectrical power transmission methods would be advisable in order to avoid the possibility of sparks. In a radioactive environment, avoidance of power transmission methods that are based on hydraulics would be advisable, because this method would pose the additional problem of cleaning up contaminated oil in the event of an accident.

4.2.3.1 Structural Elements

Three major aspects of the mechanical connection between the end effector and the manipulator are the strength and compliance of the wrist socket and the protection that the socket affords against excessive forces on the end effector. Wrist socket is used to refer to the portion of the manipulator that comes in contact with the end effector.

4.2.3.1.1 Strength - The wrist socket must be able not only to support the weight of the end effector but also to withstand any inertial forces arising from rapid accelerations. If the end effector happens to be a gripper, then the mass of any object that it carries must also be included in calculating loads.

4.2.3.1.2 Compliance - Most manipulator wrist sockets are rigid structures that provide no compliance at all. In some applications, however, much of the manipulator's work may require some form of accommodation in response to forces arising from contact with solid objects. In such cases, it may be economically advantageous to provide an appropriate compliance in the wrist socket that will either aid in performing the required accommodation motions or will perform them automatically. The best kind of compliance and the appropriate way to obtain it will certainly differ from application to application. Some possible advantages of placing the compliance in the wrist socket include the opportunity to use it with many different end effectors and a reduced moment of inertia in the manipulator. This reduced moment of inertia may make higher accelerations possible and improve production rates.

The Charles S. Draper Laboratories (CSDL) have developed an inexpensive passive device, the remote center compliance (RCC) tool, that speeds up insertion tasks remarkably. This tool is a solid passive device with unusual compliance properties. When mounted between an end effector and the object to be inserted, the tool allows the object to comply in response to forces arising from contact with the hole. This compliance makes the object enter the hole without jamming. The design and principle of operating the device are clearly explained in CSDL's reports.

4.2.3.1.3 Overload Protection - Having the wrist socket provide breakaway protection for the end effector is extremely desirable. Excessive force on the end effector should cause the following two actions to occur: (1) the mechanical connection should become compliant and (2) sensor(s) in the wrist socket should signal the work-station control computer that an unexpected exception condition has occurred. That computer should immediately take action to prevent damage to the manipulator.

Many different designs for breakaway wrists have been developed on the basis of a variety of simple mechanisms, such as

- Mechanical Fuses - These are cheap, replaceable, structural elements, such as shear pins that break or thin-walled tubes that buckle under excessive stress. Honeycomb structures are also good fuses.

- Detents - These consist of two or more structural elements that are held rigidly in position with respect to one another by spring-loaded detent mechanisms. For example, in a design by John Hill of SRI International, a disc fits into a cylindrical tube, and inward-facing ball detents at three places around the cylinder wall mate with matching depressions in the rim of the disc.

- Preloaded Springs - In these mountings, one or more pairs of structural elements are held in contact by springs (Reference 54). A force or torque acting in any direction on the end effector will tend to separate one or more of these pairs of elements in order to provide the breakaway action. The spring force establishes the level of stress on the end effector at which the breakaway action will occur.

Preloaded-spring mountings are the most desirable because they will reset themselves automatically when the force on the end effector is removed. The ball-detent mountings are the next most convenient because they require manual intervention to reassemble the structural elements. The least convenient to use are the mechanical fuses because they require not only manual intervention but also replacement of the used fuse.

The breakaway action should not leave the end effector unsupported. Mounting methods based on mechanical fuses and detents usually allow the end effector to fall a short distance and dangle from the wrist socket; this can be dangerous. Attaching the end effector to the wrist socket by a steel safety cable may or may not be advisable. One must decide whether the damage that could result from a dropped or thrown end effector would be worse than the damage that could result in its swinging from such a tether.

One must also ask what would happen if the breakaway action does not provide sufficient compliance to avoid damage. For example, if the end effector becomes stuck in a workpiece that is being carried by a powerful conveyor and the wrist socket should fail in such a way that the various power and signal connections are not damaged on the manipulator side. The end effector should be sacrificed in order to save the manipulator's wrist socket. Replacing a broken tool will not take the manipulator out of production for as long as it would take to repair its wrist socket.

A point that is often overlooked in designing a breakaway mounting is that it should break away in response to any single pure force or torque above a certain level on the end effector. To verify this, displace the end effector in any direction without rotating it and check to ensure that it breaks away. Pick an arbitrary point on or in the end effector and rotate it slightly about that point without allowing the center of rotation to move. If it still breaks away, then the end effector is fully protected.

For example, in the SRI disc-in-cylinder detent-style mounting described above, the end effector cannot be moved parallel to the plane of the disc without being rotated also. Therefore, this mounting does not provide complete protection.

In aerospace manufacturing, there will be little need for the work station to operate upon workpieces in motion on a conveyor. Therefore, the proper response to the breakaway exception condition is to simply stop the manipulator as quickly as possible. A large manipulator can be easily damaged by making it stop too quickly because of the excessive stress that the deceleration can place on components, such as gears and hydraulic lines. This possibility should be discussed beforehand with the manufacturer in order to avoid violation of warranty and service contract conditions.

4.2.3.2 Power Transmission

Most end effectors used in the aerospace industry will require power for operation. Power lines that dangle from the end effector can easily catch on equipment, be severed, and present severe hazards. Placing power conduits along (or better yet, through) the links of the manipulator is much safer. Each power line must terminate in some sort of connector at the wrist socket. The following subsections discuss connector options for various types of power flows.

4.2.3.2.1 Electrical - Electrical connections can be made through standard heavy-duty terminals if the mechanical design of the wrist socket enforces accurate positioning of the two halves of the connector. An alternative approach is to place exposed, compliant, conductive material on one or both sides of the interface. Connection of an end effector to the wrist socket then presses these terminals together. Appropriate materials include woven-wire buttons and electrically conductive, silver-doped rubber.

4.2.3.2.2 Pneumatic - Adequate pneumatic connections for either pressure or vacuum can be made by pressing two metal surfaces together with an O-ring as a gasket. Commercial pneumatic connectors may also be used if they can make and break connections reliably when simply pushed together or pulled apart.

In designing the pneumatic interface, if the end effector does not require an air or vacuum supply, the connector should plug that supply line to prevent leakage. A small air or vacuum accumulator tank at the wrist can permit use of a smaller-diameter pneumatic supply line while it provides adequate short-duration flow capacity. Regulators at the wrist can provide multiple air pressures with a single supply line. An example of a pneumatic interface is shown in Figure 13.

4.2.3.2.3 Hydraulic - Hydraulic connectors are more difficult to implement. Again, connectors that can be operated by a simple push or pull should be used. If the manipulator is hydraulic, attempting to use its own fluid supply to operate end effectors is inadvisable because of the danger of contaminating it with grit. Grit in the manipulator's hydraulic fluid can cause a servo valve to stick and can result in a sudden, rapid, unpredictable, and dangerous manipulator motion. Although advanced manipulators monitor for such events and can shut down when they occur, prevention is still better.

4.2.3.2.4 Optical - Power can be transmitted optically. One method is to simply shine light on a photovoltaic cell array on the end-effector side. The light need not be coherent. Because of the difficulty of transmitting much power this way, this method is applicable only in special situations, such as when operating in an explosive atmosphere. The same beam of light, however, can also carry information, and this method may have advantages in some situations.

4.2.3.2.5 Mechanical - Power can also be transmitted mechanically. For example, a motor on the manipulator side can rotate a splined shaft which mates with a shaft on the end-effector side. This method can reduce tooling costs by allowing one motor to be shared between several end effectors.

Transmitting the shaft rotation through a flexible cable from a motor that is mounted farther back along the manipulator can improve performance by reducing the mass and weight at the wrist.

If the manufacturing application demands it, limited amounts of mechanical power can be transmitted to a hermetically sealed tool through flexible elastomeric or metallic membranes. A variety of drive mechanisms for this purpose are commercially available (such as bellows, peristaltic plates, and wobble drives).

4.2.3.3 Information Transfer

Most aerospace end effectors will require control information from the workstation computer, produce information for it to use, or both. Information is usually transmitted at low power levels. This procedure makes it easier to design connectors and conduits, but it also introduces the problem of noise susceptibility.

Several different information flows may be multiplexed into a single signal channel. Multiplexing may be advisable if the cost of multiple connectors is too high or if their overall reliability is too low. The following sections discuss various connection options for transmission of information to and from the end effector.

4.2.3.3.1 Electrical - The easiest way to transmit information is electrically. A wide variety of commercial electrical connectors is adequate for use in interfaces. The major cause of unreliability will be bad electrical contact between the mating conductors due to oxidation or contamination by dust and oil. Many commercial connectors are designed specifically to mate reliably under these conditions (e.g., by mating with a wiping action that scrapes away contaminants).

Figure 13
PNEUMATIC POWER TRANSMISSION

4.2.3.3.2 Pneumatic - Many commercial control and sensing elements transmit and receive information via a 3-15 psi (0.2-1.1 kg per sq cm) pneumatic signal. Connectors appropriate for transmission of pneumatic power should also be able to make connections for pneumatic information transmission.

4.2.3.3.3 Optical - Modulated light can carry extremely high bandwidth signals through very small fiber optic cables with no noise pickup problems. The work-station control computer could not possibly process that much data, however, so the bandwidth capability is probably not significant. Noise immunity is a more important feature of optical data transmission. Present-day commercial fiber-optic cable connectors are not rugged enough to serve in a wrist socket, however. Instead, the light signal should be transmitted through a small air gap in the interface from a modulated light-emitting diode to a photocell.

A coherent fiber optic bundle can be used to carry an optical image from a lens in the end effector to a television camera in the wrist socket. This process would allow many end effectors to be equipped with vision capabilities at a very low unit cost. The camera should look directly across the interface gap at the near end of the bundle. Split bundles can combine multiple points of view in different regions of the camera image.

For illumination of the workpiece, a lamp in the wrist socket can direct a beam of light into the near end of the bundle. The same bundle can be used for both illumination and image acquisition by placing a half-silvered mirror in front of the camera so that the camera looks directly through it while light from the lamp is reflected off the mirror into the near end of the bundle. While this arrangement tends to minimize the number of shadows in the scene, it may emphasize specular reflections from surfaces normal to the line of sight.

4.3 FIXTURES AND TOOL ACCESSORIES

An aerospace manufacturing work station will include many more kinds of tooling than just end effectors. The work station may include tooling for calibration, measurement of tool wear, jigging workpieces and templates, feeding and orientation of small parts, and brush tables.

Some general principles to keep in mind when designing auxiliary tooling are

- Provide access to the tooling for maintenance personnel
- Protect cabling on the floor from objects that the manipulator may drop on them from above
- Either design them to be sturdy enough to withstand the maximum force that the manipulator can exert or equip them with breakaway mountings as described in Subsection 4.2.3.1.3.

4.3.1 Templates

Templates are one of the most important kinds of tooling that will be used in the work station until templateless machining techniques are perfected. Two kinds of templates most often used in aerospace manufacturing are (1) fiberglass layups containing bushings for robotic hole drilling and (2) perforated sheet metal panels for guiding manual semiautomated drilling operations (References 51-53). If the work-station controller at some point takes over the job of jigging templates in place on manual semiautomated drilling operations (References 51-53). If the work-station controller at some point takes over the job of jigging templates in place on workpieces, then it becomes necessary for the controller to be able to identify templates in order to ensure that it is using the correct one. A variety of methods may be used to mark the templates in a machine-readable way, such as OCR characters, bar codes, perforation or notch patterns in its surface or edge, and patterns of embedded permanent magnets.

4.3.2 Tool Storage

In any work station where a manipulator uses a multiplicity of end effectors and other tools, storage must be provided for the tools that are not being used. The manipulator should be able to pick up an end effector from the tool storage area and put it back without manual assistance. Proper design of the manipulator's wrist socket will permit this.

It is desirable for the work-station controller to be able to distinguish one end effector from another in some way because, for example, a human operator could accidentally place the wrong end effector in a tool rack. If a mishap during a production run should make it necessary to restart the work-station control computer, time will be saved and possible additional problems will be avoided if the computer can determine automatically whether or not the manipulator is holding an end effector and which one it is. Some ways of identifying an end effector include

- A binary-encoded tool number readable through some of the wires in an electrical information connector in the wrist socket
- A binary-encoded tool number in a pattern of small permanent magnets that can activate magnetic switches
- A bar code
- A unique shape that can be recognized by the vision software
- A unique weight that can be read by a force-sensing wrist.

Ideally, the end effector identification information should be available to the work-station control computer through the wrist socket. If an unidentified end effector has to be carried to a reading station in the work area for identification, there is the additional problem of having to know what its shape is before starting to move it, so as not to hit anything with it on the way.

Providing a wear sensor for drills, routers, grinders, and similar tools may be worthwhile. The tool storage area may be a convenient place to locate it. A sensor in each slot in the tool rack will save production time; if the sensor is expensive, there should be only one.

4.3.3 Jigs

Jigs are a major expense in aerospace manufacturing. These heavy bulky objects must often be stored between production runs of a given aircraft. With present-day manual fabrication techniques, the need to move one of these large jigs is only occasional. Cranes and manpower easily satisfy these needs now. In automated aerospace plants of the future, processing times at each station may shrink by an order of magnitude. This shrinkage will aggravate material flow problems, and management should be alert to the possibility of identifying work centers in which automatic transport mechanisms would be cost-effective.

Current jigs position each part of an assembly accurately with respect to the other parts for fastening operations, such as drilling, countersinking, and riveting. For use with conventional industrial manipulators; however, the jig must also position the assembly accurately with respect to the manipulator because currently available commercial manipulator control software provides only very limited indexing capabilities for adaptation to an arbitrary workpiece position. This adaptation process is called automatic indexing (Subsection 5.2.1.5). If available manipulator control software should be improved to support automatic indexing, the assembly would not have to be positioned accurately. This improvement would present an opportunity for cost savings in tooling (Reference 55). However, the assembly must still be held in place rigidly enough to withstand any contact forces that may arise during operations.

4.3.4 Other

Some amount of tooling will be required for other activities such as calibration, tool-wear monitoring, and part orientation and presentation.

4.3.4.1 Calibration

Calibration of various sensors will probably be a frequent activity in the work station. Calibration activities will require equipment such as the following:

- For Vision - Objects of known size for calibrating lenses, reference marks for determining the position of a camera in the work area, and the position of the camera relative to the wrist socket of the manipulator

- For Force-Sensing: Weights and pulleys for exerting known forces or torques on the sensor

- For Proximity Sensors - Surfaces whose signal-reflectance properties are known for determining curves of sensor signal versus distance

- For Tool Sizing - Fiduciary marks and lines in known positions that can be used by a human operator in placing end effectors in known positions in order to enable the work station to determine their size and shape from the position of the wrist socket.

If there is frequent need for a given calibration procedure, the work-station controller should be able to perform the procedure automatically. Otherwise, a manual procedure will probably be better because the people involved will have an opportunity to inspect the production equipment closely and may be able to detect wear or incipient failures that would otherwise have gone undetected.

In principle, every sensor in the work station requires some sort of calibration (except perhaps for some of the simple binary sensors such as microswitches). This includes the joint-position sensors in the manipulator itself. It would be advantageous if the work-station controller could calibrate all these sensors automatically.

Some of the more complex calibration procedures that may be required include

- o Generating a table of positioning errors for the manipulator
- o Measuring the position and orientation of a camera, as well as the magnification factor and field of view of its lens
- o Determining the sensitivity and offset readings of a force- or torque-sensing wrist or worktable
- o Locating a jig or workpiece (indexing).

4.3.4.2 Tool Wear

In order to maximize productivity, tools should be changed as infrequently as possible. Various methods, such as testing for the presence or absence of a tiny pellet of irradiated Tungsten at the wear limit of the tool (by sensing its radioactive emissions), have been developed for sensing tool wear electrically (Reference 56). Vibration levels and torque are other indicators of tool wear, but using them requires extensive calibration measurements for a given tool and workpiece material. Optical measurement of tool wear is difficult to do on-line because of complex tool geometry and the small dimensions involved.

4.3.4.3 Generalized Jigs

Jigging is one of the major expenses in aerospace manufacturing. Having a few, expensive, computer-controlled, general-purpose jigs rather than building many special-purpose jigs may prove to be cost-effective. Alternatively, group-technology studies may suggest designs for jigs that can hold a variety of different parts with minor adjustments. Including in any jigs some marks or tooling points that the work-station controller can sense in some way will probably be useful. This will aid in automatic calibration and indexing.

4.3.4.4 Part Orienters

Small parts can be oriented fairly well by vibratory feeders with specially designed tooling for each part. Orienting larger parts requires different methods because the energy required to orient them with vibration is excessive. Noise levels become dangerous, and the parts themselves can be damaged by colliding with each other. The work station will probably require automatic rivet feeders for rivets and other fasteners.

SRI International has developed a prototype, general-purpose, microprocessor-driven device that uses vision to detect the orientation of a part and then pushes the part over the edge of a step in order to change its orientation (Reference 57). This device is suitable for parts that are too large to orient by conventional vibratory feeder methods. Such devices may become commercially available in the near future. At present, however, the aerospace industry is mainly concerned with automation of sheet metal processing in which orienting workpieces is a small part of the problem. As experience is gained, management may want to reap the cost benefits of automation in other manufacturing areas that process irregularly shaped objects weighing a few pounds and up to a foot in size, such as small forgings and bracket assemblies. Retaining orientation of such parts in certain processing and transporting operations, such as cleaning baths, is often uneconomic. A need for tooling that can orient batches of fifty or a hundred of these parts quickly and cheaply may then arise.

4.3.4.5 Part Presenters

Much aerospace manufacturing involves the application of fasteners, and these fasteners will have to be fed to the appropriate tools for insertion. The feeder equipment will be an important part of the work-station tooling and should be selected upon the basis of its reliability. One of the most troublesome problems in the automatic feeding of small parts is that they jam in the feeder mechanisms. This jamming is not so serious in manual assembly, for people are so dexterous that they can clear those jams very quickly. Unfortunately, programming a work station to correct any kind of part-feeding jam that might occur will not be practical simply because of the enormous variety of failure modes. Two solutions may be adopted to increase the reliability of the part-feeding system. One approach is redundancy; another is 100 percent inspection of small parts before attempting to feed them. Redundancy is only practical for small subsystems of the feeder system. For example, one bowl feeder might be equipped with two output tracks, each equipped with a jam sensor. Whenever one jammed, the work station would immediately start taking track as quickly as possible. Inspection of all small parts is now a more likely possibility than it has been because of the appearance of a wide variety of low-cost, microprocessor-based visual inspection systems in the marketplace. It is also practical for a manufacturer to put together in-house his own inspection system optimized for the particular class of fasteners that he uses most often. One of the simplest and most effective screening methods is to simply weigh each small part since a large proportion of defects tend to result in addition or deletion of material.

Hill and Park of SRI International have demonstrated a programmable bowl feeder in which an SRI vision module replaces fixed tooling in the part feeder track.

The McDonnell-Douglas Corporation (MDC) has pointed out the utility of brush tables for part presentation in sheet metal processing (Reference 58). A brush table is a table whose upper surface is covered with upward-pointing bristles. The main advantage with the brush table is that a gripper can easily pick up a piece of sheet metal that is lying on such a table. The bristles will part to allow the lower claw or finger of the gripper to pass under the sheet so that the gripper can hold the sheet by one edge. The tables should have useful characteristics such as backgrounds against which to sense parts visually. MDC has generated a number of promising design concepts for work-station tooling based on brush tables.

-5-

Work Station Integration

Automated industrial systems have always required relatively advanced engineering and technical skills. Robots that utilize computer control are programmable, and possess computer-enhanced adaptability have not lessened this requirement. In fact, the technical areas associated with robotics have multiplied not only in quantity but also in complexity. For this reason, the potential user of robots must establish a well-organized approach to system design.

This section is a guide for system design and discusses programming, control functions, and control structures.

5.1 PROGRAMMING THE WORK STATION

Programming means the generation of algorithms and data. An algorithm is a description of a sequence of actions. Automated manufacturing will involve literally hundreds of computer programs, large and small, interactive, batch-oriented, and real-time, that must operate in harmony with one another. (Aerospace manufacturers will be well-advised to pay as much attention to standards in the programming shop as they do in the machine shop.)

It is both practical and possible for programs to generate other programs -- not from scratch but from abstract high-level descriptions of what the generated programs should accomplish. The decisions on how the generated programs should accomplish the goal are made by the generating programs faster and better than people can. Translator programs for programming languages such as APT, FORTRAN, COBOL, and PL/I, for example, do exactly this (Reference 59). The new ADA language developed by the DoD should be very important in future robotic system programming (References 60,61).

In the future, task programs will probably be generated more from linguistic descriptions (initially, in formal programming languages) and less from interactions with a human trainer (References 62-64). LAMA is an ambitious robot programming language of this kind whose development was begun at MIT. Some of the more advanced robots available today are already beginning to follow this trend. Some allow interactive training of conditional branches within the manipulator-motion sequence, and others permit the trainer to type a program in a language similar to the popular BASIC programming language for very complex manipulation tasks (Reference 65). A number of advanced robot control languages are now in use in laboratories, such as WAVE and AL at Stanford, AUTOPASS at IBM, and PAL at Purdue (References 10,66-68). Park has made a survey of many other robot programming systems (Reference 69).

The flexibility of a robot programming system depends upon the basic operations which the robotic system can perform, the available control structures to specify how and when to perform the operations, and the facilities provided for development of robotic programs. In order to understand how these system characteristics come about, the different levels of software that are involved must be understood.

5.1.1 Programming Levels

At least three distinct types of robotic programming are necessary, and each type requires a distinctly different level of programming skill from the person doing it. In large factories of the future with many robots, these three types of programming will very likely be done by different people, and three different job classifications may well come to be recognized. These three job classifications are the system programmer, the task programmer, and the equipment operator.

5.1.1.1 System Programmer

The system programmer would write, debug, and maintain all the robot-control software which will be generally useful in a variety of tasks a given robotic system might be called upon to perform. This job classification requires the highest skill level of the three, and requires an experienced computer professional. This programmer would very likely work closely with other professionals, such as producibility engineer or manufacturing engineer, who are skilled in the particulars of the relevant manufacturing processes. The system programmer would provide a set of interactive programming procedures for the task programmer to use. These procedures must be easy to learn and apply and should be generally useful in controlling the automatic equipment in the manufacturing cell. Most of all, these procedures should be extremely tolerant of mistakes which the task programmer may make in using them, and they should prevent his mistakes from resulting in damage to the equipment or danger to the people working near the robot.

5.1.1.2 Task Programmer

The task programmer would use the programming methods supplied by the system programmer in order to create task programs and descriptions of algorithms (and any required data items) that make the robot perform specific industrial tasks. The task programmer must be familiar with the available programming methods, the capabilities of the automatic equipment, and the requirements for manufacture of the product. If the task programming methods are properly designed, the task programmer should not need much or any prior computer programming experience, although programming knowledge will help produce more efficient applications programs.

5.1.1.3 Equipment Operator

The equipment operator would start up the robotic equipment at the beginning of the workday, come to its rescue when it gets into trouble, and shut it down at the end of the shift. For simple tasks, the equipment operator will not have to give the robot any particular information. All he will have to do is ensure that its tools are all available and in working order, that its work area is clear of obstructions, and that it gets a steady supply of workpieces to process. However, in more complicated industrial tasks, the robotic system will need some small amounts of information, perhaps at rather frequent intervals - every shift or every hour. The person who can supply that information most conveniently will be the equipment operator. In principle, an appropriate sensor could obtain almost any kind of information that the robot might need; in practice, there will always be last-minute complications in a production run that have to be taken care of quickly and inexpensively. The equipment operator will be the best man for the job.

5.1.2 Basic Program Functions

Basic functions are those discrete unit actions that the task programmer can direct the robotic system to perform, such as moving a tool to a specified position, operating an end effector, or reading a number from a sensor or manual input device. The system programmer would actually be responsible for choosing the set of basic functions that would be most useful to the task programmer in his work. The system programmer has to use his skills to identify a small set of rudimentary activities which comprise the whole class of manufacturing tasks likely to be performed by a given robotic work station. The system programmer would then invent a methodology by which the task programmer could quickly and easily string together sequences of these activities to perform any task in that class. One convenient categorization of basic functions is

(1) Computations

(2) Decisions

(3) Communications

(4) Manipulator Movements

(5) Tool Commands

(6) Sensor Data Processing.

Most present-day robotic systems offer only manipulator movements and tool commands and a simple form of sensor data processing, such as sensing relay closures. Some commercial robots which do provide all six types of basic operations include Unimation's PUMA arm which uses the VAL language; Olivetti's two-armed SIGMA robot, which uses the SIGLA language; and IBM's manipulator -currently available only within IBM - which is programmed in the EMILY language.

5.1.2.1 Computations

The ability to specify computations to be performed during a task is one of the most important capabilities to include in a robot control system. Without it, the task programmer must hope that the system programmer had enough foresight and detailed understanding of actual factory operations to provide software capable of handling every possible contingency; this is optimistic.

If the robot is not equipped with any sensors there is probably no need to specify computations in a robot program. Without sensors, a robot is nothing more than a numerically controlled machine suitable for programming in a language such as APT. All it can do is wave its tools around in space and hope that the workpieces and jigs are exactly where they are supposed to be.

Some of the most useful kinds of computation that a sensor-equipped robot can make are analytic geometry calculations. These computations can enable the robot to decide for itself where it has to put its tool or gripper next. Since the exact computations needed for a particular industrial task are usually highly specific to that task, it is impractical for the system programmer to provide canned procedures to cover every possible circumstance. The best that he can do is to provide the task programmer with a complete set of computational tools to cover the unusual situations. It may well be that only the more experienced task programmers will be able to use them effectively though. This situation is similar to that in which a skilled machinist will make a special-purpose jig for an inspection procedure that he will have to repeat many times; but he will fall back on his general-purpose but more difficult to use micrometers and dial gauges when an occasional part requires a different measurement.

A useful but neither vital nor all-encompassing set of computational tools for analytic geometry calculations might include the following:

- The arm solution and back solution
- Operations on coordinate frames and position representations (e.g., composition of relative positions, change of coordinates)
- Vector operations (e.g., dot product, cross product, length, unit vectors, scaling, linear combination of vectors).

Naturally, these calculations would go along with a full set of arithmetic operations, the square root function, and the trigonometric functions. (Incidentally, the arc tangent function of two arguments turns out to be used much more often than the arc sine or arc cosine functions).

5.1.2.2 Decisions

A robot control system can make decisions based on sensor inputs without performing any computations, but the ability to make decisions about what to do next based on results computed from raw sensor data (as well as stored data) really makes a robot control system powerful. A single conditional branch instruction (say, a test for a zero value) would be sufficient to implement any decision algorithm because the result of a computation can always be put in such a form if the task programmer tries hard enough. However, the task programmer's job is much easier if he has many different types of conditional branches to choose from. Some useful types to have available include sign tests (positive, zero, or negative) and relational tests (greater than, not equal to, etc.), Boolean tests (ON or OFF, TRUE or FALSE), logical tests (testing groups of bits in a computer word), and set tests (member of a set, emptiness of a set, etc.).

5.1.2.3 Communication

The ability to communicate with the operator allows the robot to ask the operator for information, to tell the operator what he ought to do next, and to let the operator know what it intends to do. A person and a machine can communicate in many different ways. Some ways are very simple procedures requiring only simple equipment, while others are very sophisticated and require expensive electronics. In order of increasing sophistication, some of the possible output devices through which the robot can present information to the person include

- Back-lit messages which the robot can display by turning on its lamps
- A character printer, display screen, or any of a wide variety of character display devices based on arrays of light-emitting diodes, plasma cells, electroluminescent panels, incandescent wires, or liquid crystals. Some of these are extremely bright and legible, even from twenty feet away.
- A plotter or graphic display screen
- A speech synthesizer or other audible device (bell, horn, etc.)

Some devices which enable the person to say things to the robot include

- Pushbuttons, toggles, knobs, and thumbwheel switches

- o A numeric or alphanumeric keyboard
- o A light pen, track ball, digitizer table, Rand tablet, or SRI (Stanford Research Institute) mouse
- o A teleoperator master control (teach gun, pendant, etc.)
- o An optical character reader (OCR).

The simpler the devices, the lower the skill level required of the equipment operator. Speech input-output devices are certainly not simple, but they may prove to be an exception to this rule. At present though, their capabilities are still extremely limited (References 70,71). Whether even an excellent speech input-output device would be of much use in a real factory situation with personnel of very low skill level is not yet clear.

5.1.2.4 Manipulator Movements

Manipulator movements can be described in many different ways. Historically, the first industrial robots allowed the robot's programmer to specify only a sequence of point-to-point motions, with each point being described in terms of a set of manipulator joint positions. The manipulator would stop at each trajectory end point and perhaps wait for an external signal before going on to the next point. Via points permitted greater control over the arm trajectory by allowing the programmer to specify points through which the arm should pass without stopping. Interpolating many intermediate set points between a pair of programmed joint-position set points reduced the number of tool positions that had to be trained for a close-tolerance path-following application such as arc welding. Complete motion sequences could be selected and performed on the basis of an external signal. All of these capabilities were first made available in robot arms that had no computers.

Adding a computer to a manipulator greatly increased its usefulness in the following ways:

- o Much more complex motion sequences become possible.
- o Sensor-controlled manipulator motions become possible.
- o The position of the tool could be stored in a format that was independent of the design of the manipulator and of its calibration factors.

An actual arm movement can be specified in a number of different ways. The simplest way is to give a set of joint positions to the joint servos and just wait until the servos arrive at those positions. A more sophisticated way is to interpolate arm positions in joint space. This procedure makes all the joints start moving and stop moving at the same time. Giving tool positions in terms of coordinates that are not related to the shape of the manipulator is even more advanced and (except for X-Y-Z arms) requires a computer for arm solution computation. Interpolating tool position in Cartesian space produces smooth motion of the tool tip along a trajectory for path following (References 23,72,73). Introducing a reference frame in which to describe tool positions and then allowing that frame to move is convenient in many situations. The frame may have two or more fixed locations in the workspace in order to represent several identical work stations. Alternatively, the location and orientation of this frame might be made to vary in some way with time or to depend upon the instantaneous value of some sensor reading, such as a position transducer on a conveyor to track the workpieces that it carries.

Aside from the measuring system used, motions can also be described as being absolute or relative. An absolute motion carries the tool to the same position in the workspace every time, regardless of where the tool comes from. A relative motion moves it a specified amount from its initial position. Where the tool goes depends on where it comes from. A motion subroutine using only relative motions may be defined in the sense that you can make the last relative movement bring the tool back to its starting position.

Relative motions are usually more trouble to use, however. A sequence of relative motions may be trained with no problems by starting with the tool in one position. When you try to play the sequence back starting with the tool in a different position, you may find that one of the joints will hit one of its limit stops.

Four-by-four Denavit-Hartenberg matrices (References 21-23) are a very convenient way to represent the position of a reference frame as well as the position of an object relative to a frame and the shape of a tool or gripper. They can be multiplied together to determine the location of an object "A," whose position is specified relative to a second object "B," whose position is in turn given relative to a third object "C," and so on. This composition of relative position operators is also a simple way to compute the back solution (the tool location, given the joint positions) for any existing industrial manipulator.

The proper choice of an external coordinate system or reference frame in which to represent positions of objects can sometimes simplify manual training procedures where the tool must be positioned precisely by remote control. The most common system used is the Cartesian or X-Y-Z system. Cylindrical and spherical coordinate systems have not proven as useful, perhaps because they are always centered on the robot. Such systems might be more useful if the task programmer could specify their location and orientation in order to align them with the surrounding equipment instead.

5.1.2.5 Tool Commands

A tool control command is generally initiated following a switch or relay closure. The relay may, by switching electric power on and off, control the tool directly, or it can send a low-power signal to an electronic controller that actually operates the tool. Direct control is the simplest method and requires little from the robot control system. Other sensors can be used to sense progress and completion of the tool functions.

By using a tool function controller, whether internal or external to the main robot controller, more sophisticated control is possible. With this type of system, the robot controller positions the manipulator and communicates with the tool function controller. When a tool function is initiated by a sensory device, control is transferred to an internal subroutine or to an external controller. Operations of the tool functions are then carried out by the tool function control system; upon completion, control is returned to the robot controller. With separate control systems, tool function control and robot control can operate concurrently if the operations do not conflict and if control interaction conflicts (deadlocks) are compensated for. Control transfer and concurrent control methods have been successful in airframe panel drilling and routing applications.

5.1.2.6 Sensor Data Processing

Sensors are the most important new development in industrial robotics. The full utility of a general-purpose computer used for manipulator control cannot be realized until it is connected to sensors. Sensors come in a bewildering variety of forms. Sensors may be categorized on a functional basis as follows:

- Proprioceptors sense the position of the arm or other computer-controlled articulated mechanisms.
- Touch sensors sense physical contact between the tool and another object.
- Proximity or range sensors sense distance from the tool to a workpiece or obstruction.
- Force and moment sensors sense the forces and moments that arise during fitting operations such as insertions.
- Visual sensors "see" the objects in the workspace in order to locate and identify them.

Proprioceptors are usually just the position feedback transducers on the individual joints. Much can be done with a manipulator, even if the signals from these sensors are not available to the controlling computer. The Cincinnati Milacron T3 arm, for example, never computes a back solution to find out where the tool is. Instead, it keeps in memory the last Cartesian position to which it sent the tool and assumes that it got there. Nevertheless, this information is sufficient to allow the T3 control computer to make specified movements relative to the current position in Cartesian coordinates and to move the arm along or around a set of X-Y-Z axes passing through the tool tip.

Touch sensors are usually microswitches, although any transducer which responds to the close presence of an object could also be used. Reed switches can be used to detect the presence of specific objects in which permanent magnets have been imbedded. A touch sensor can be made very compliant by attaching a long spring whisker to it. This allows sufficient time to stop the arm after the whisker touches something. However, if the arm moves quickly, the inertia of the whisker may cause false alarms.

Some physical effects often used in proximity detectors include infrared reflectance, ultrasonic sonar, and eddy-current induction. Their useful operating range varies from about 0.03 to 4 inches (0.1 cm to 10 cm). Longer distance measurements require a range-measuring device.

Force and moment sensors are useful in fitting, fastening, and packing operations for monitoring contact forces between an object carried by the hand and a stationary object. These sensors can also be useful in inspection or identification of workpieces when they are used as a sensitive scale to measure the weight or mass distribution of a workpiece. Measuring the applied force and torque about a point remote from the sensor is practical so that the sensor can be mounted in the manipulator's wrist or in the work table yet be able to measure stresses at the tool tip or at an arbitrary point or (even in) the workpiece.

Vision systems using television can be applied in a variety of ways, including visual acquisition, identification, and robot positioning. Characteristics of the type of television system that improves the computer interface are

- A square aspect ratio, rather than the standard 4:3 aspect ratio of home television, and pixels (picture elements) arranged on a square grid, rather than a rectangular one.

- A computer-compatible digital interface, rather than an analog interface carrying standard EIA video signals.

- Often, some provision for binary imagery that is easier than gray-scale imagery for a small computer to process quickly.

In addition, the light-sensitive components in the cameras themselves are generally solid-state diode arrays, rather than image orthicon or vidicon vacuum tubes. Diode array technology has several advantages over tube technology, such as reduced image distortion, no need for high voltages or complex beam deflection components, anti-blooming, light weight, small size, extreme ruggedness, and the economies of scale typical of semiconductor devices for large production runs. Cameras soon to be available will feature programmable readout in order to provide windowing, zooming, and nondestructive readout which permits in-camera image processing, such as Hadamard transformations and edge enhancement.

5.1.3 Software Design*

An important aspect of efficient computer control of robotic systems is well-designed quality software. This section describes the characteristics of high quality software and the uses of top-down design and structured programming as methods for achieving high quality software. Specific suggestions and guidelines, as well as advantages, are described.

5.1.3.1 Characteristics of High Quality Software

Quality in software is a complex issue. There has been precious little focus on quality characteristics independent of functional requirements; this tends to produce an uneven spectrum of software products. Yet the characteristics of quality in software can be addressed. A list of such characteristics is presented below for the convenience of the reader.

Correctness. Programs perform exactly and correctly all the functions expected from the specifications, if available, or else from the documentation. Incorrect documentation is as serious as an incorrect program. Correctness is an ideal quality that is rarely determinable, so a more practical quality is reliability.

Reliability. Programs perform without significant detectable errors all the functions expected from the specifications or the documentation. High reliability indicates that programs are relatively trouble free in performing what they claim to do. An equally important question is whether the functions and performance are adequate and suitable to a needed purpose. The latter quality is called validity.

Validity. Programs provide the performance, all functions, and appropriate interfaces to other software components that are sufficient for beneficial application in the

*Adapted from "Robotics Support Project For the Air Force ICAM Program," Second Quarterly Interim Report, National Bureau of Standards, April 1979.

intended user environment. The software, without additional programming or manual intervention, has the capabilities that reasonably would be expected for its purpose. Validity is a quality of specifications as well as computer programs. Examples of an invalid program would be an interactive editor that had no online function for retrieving stored text for inspection or a FORTRAN language compiler that had no DO loop implementation. Validity involves judgment of user requirements and may change if the intended application or purpose is altered. Because poor reliability may render a needed function useless, reliability is necessary to validity.

Resilience (or robustness). Programs continue to perform in a reasonable way despite violations of the assumed input and usage conventions. Input of unacceptable data or an inconsistent command should never cause a result that is astonishing and detrimental to the user -- such as the deletion of any valid results obtained previously. Programs should include routine checks and recovery possibilities that are forgiving of common user and data errors. Resilience is related to the broader quality of usability.

Usability. Programs have functions and usage techniques that are natural and convenient for people and that show good consideration of human factors and limitations. For example, the programs have few arbitrary codes for data in input or output, have consistent conventions in different operating modes, and provide thorough diagnostic messages for errors or violations of use.

Clarity. The functions and operation of the programs are easily understood from the user manual, and the program design and structure are readily apparent from the listing of program statements. This means that documentation must be well-written and that the program is carefully designed with meaningful choices of variable names, use of known algorithms, frequent and effective comments in the program to describe its operation, and a modular structure that isolates separate functions for examination.

Maintainability. Programs are well-documented by manuals and internal comments and are so well-structured that another programmer could easily repair defects or make minor improvements. Clarity is essential for maintainability. Also implied are a wide variety of good design attributes, such as program functions that help to diagnose potential problems, e.g., periodic reports of status or control totals or general techniques that can be readily adapted for change, e.g., the isolation of constants, report titles, and other static data as named variables.

Modifiability. Program functions that might require major change are well-documented and isolated in distinct modules. Maintainability is essential to modifiability, but modifiability means that a concerted effort was made to anticipate major changes and to plan the software design so that they could be made easily.

Generality. Programs perform their functions over a wide range of input values and usage modes. Programs are not limited to special cases or ranges of values when the functions are commonly or reasonably extendable to a more general case.

Portability. Programs are easily installed on another computer or under another operating system. A standard programming language is used, and hardware or other software-dependent features are isolated for easy change.

Testability. Programs are simply structured and use general algorithms to facilitate step-by-step testing of all capabilities.

The terms and definitions presented above can be thought of as guiding principles. With the use of these principles and those of the following subsections, top-down design and structured programming, quality software can be produced.

5.1.3.2 Top-Down Design

The top-down design approach has as its goal the logical development of a system design that can be implemented by the technique of structured programming. Top-down design is a design philosophy that evolved in response to the need for the production of large reliable software programs. The concepts of top-down design are sometimes called stepwise refinement, hierarchical design, constructive programming, and a variety of other names.

The key concepts in this design approach are to develop a functional description of the system and to identify the inputs and outputs of the functions. The design proceeds in stages. At the first level or top of the design, the whole system is defined as a single function.

The next level of the design is achieved by a more detailed breakdown of the function in the level above. In general, a function box is replaced by two to six smaller component functions, which taken as a whole represent the complete function at the higher level. In Figure 14, a single function which might correspond to the top level of a design is depicted. In Figure 15, the single function of level 1A has been broken down into three separate functions, each of which represents a part of the original function. The important principle is that as one creates each new level of the design, the functional elements in the level above are refined, and more detail is entered into the design. At each lower level in the design, the details of the system are more apparent, and each level of the design represents a complete view of the system.

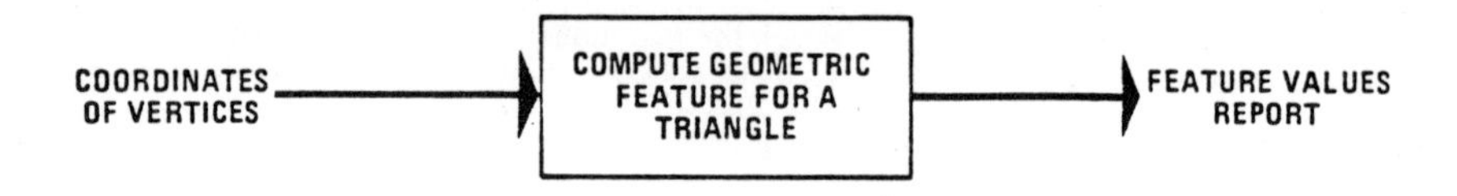

Figure 14
TOP LEVEL DESIGN

This approach provides an overall view of the system at many levels. Thus design reviews can be conducted by a wide variety of individuals. The user can explore the design at the levels compatible with his expertise, as can a manager, a system analyst, and a programmer.

Typically, the higher levels of the design are machine independent, and it is only at the lowest levels that the functional characteristics of hardware must be addressed. This concept can also be applied to top-down design.

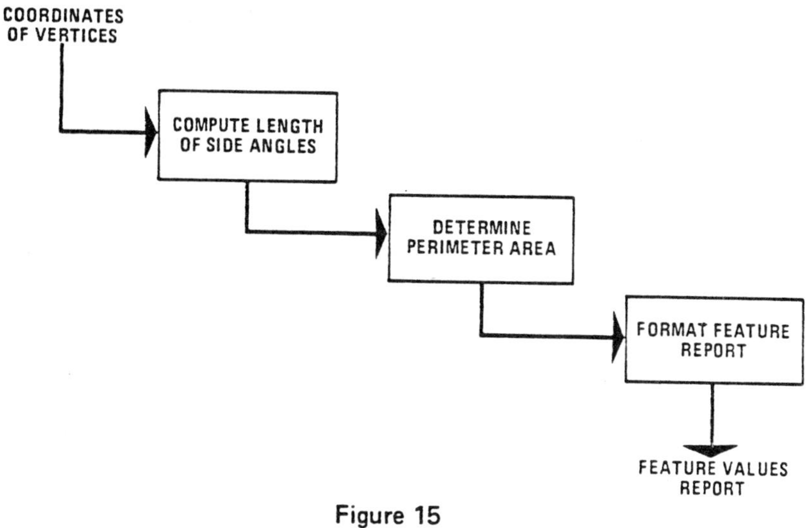

Figure 15
LOWER LEVEL DESIGN

Several suggestions for successful top-down design are presented and explained in the following paragraphs.

1. For each function defined, provide a precise description of the inputs and outputs. A function description is based on the transformation of a set of inputs to outputs. The clarity of the design requires that all three elements function and that input and output be well understood.

2. Limit the size of the expansion to what can fit on one page. The rationale behind top-down design is to gradually refine a global system description to the details of a system design. In order to maintain the orderly decomposition of the global to detailed description, the reader (and often the author) must be able to absorb the details of the functional decomposition slowly (a page at a time).

3. Try to ignore the details at lower levels in the design before the lower level is reached. A frequent temptation is to worry about the implementation details of the lower levels long before the higher levels have been expanded. This often leads to digressions and needless worry about problems that may not materialize. When carried out properly,

the design is iteratively refined by taking two steps forward and one step back. Rarely if ever must the design of levels several steps higher be modified due to some new detail which becomes evident at a low level.

4. Pay close attention to the design of data and data structures. The data is the interface between functional modules. As the design develops, so does the data structure. It is as important to design the data structure as the functional decomposition of the system. Don't fall into the trap of independent program and data structure design.

5. A formal mechanism for the documentation of the design is required. The exact mechanism is not as important as the existence of one and its rigorous use by all designers. Typically the design is maintained by a special librarian whose job is to keep the records, circulate the design for review, and keep track of corrections and revisions. The design is complete when each functional element has been expanded into a form appropriate for coding. Such an element is generally referred to as a module. A module should

   o Perform a simple and well-defined function

   o Have a complete description of inputs and outputs

   o Correspond to a single subroutine or procedure in a structured programming language

   o Have a one-page language code that includes up to 50 source language statements.

A top-down design has many advantages. First and foremost, this design provides a formal mechanism for breaking the design of a complex process into a coherent set of functional descriptions. Second, the design is structured to permit review at many levels as the design progresses. Third, an implementation of the design by the methods of structured programming is facilitated by the modularity of the functional elements produced in the design. Finally, systems designed in this fashion are easily modified or expanded at and below the level affected by the change. Thus a change of computer, a minimum change in requirements, or the expansion of the design is often easy to perform, and the affected software modules in the implementation are readily identified.

5.1.3.3 Structured Programming

The early work on structured programming was inspired by an attempt to create programs which can be proven to be correct (in the mathematical sense). While this goal has not been achieved in any practical form, many other benefits have been achieved. Sometimes called "Ego-less" programming, structured programming practices are designed to reduce the dependence on individual programmers and to facilitate team efforts. Furthermore, software developed in the structured discipline is easier to modify, maintain, and enhance. The five major objectives of the structured programming discipline are

(1) Program readability and clarity

(2) Increased programmer productivity

(3) Reduced testing time

(4) Reliability

(5) Maintainability.

The objectives are achieved by adhering to a structured discipline in the creation of the software. In the discussion that follows, it is assumed that a top-down design effort has been carried out to define a set of software modules. The creation of an individual module is now addressed.

Three basic building blocks are used to construct a program. Each of the three blocks has a single entry and exit. The process box (shown in Figure 16) may be thought of as a single computational statement, or as any other proper computational sequence with only one entry and one exit. Thus, a process corresponds to a well-defined computation and might invoke an entire procedure or be a single machine language instruction. The important part of the definition is the single entry and exit.

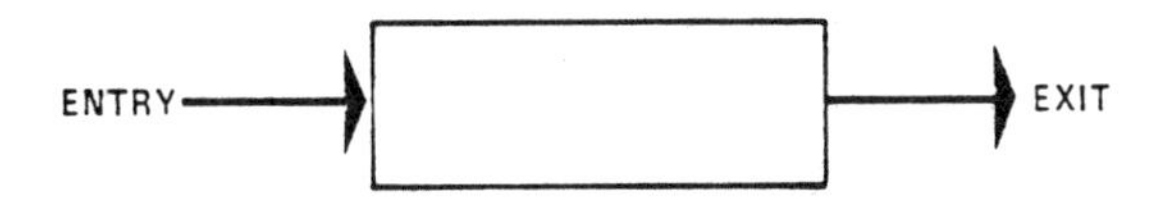

Figure 16
PROCESS BOX

A generalized loop mechanism (shown in Figure 17), usually called a DO WHILE loop, has a single entry and a single exit taken when some condition is false. This loop includes a process to be repeated while the condition remains true. At some point in time, the process (or an external event) must change the state of the condition being tested; otherwise, we would have an infinitely repeating loop.

The binary decision mechanism (shown in Figure 18), often referred to as an IF-THEN-ELSE statement, has a single entry to a TRUE-FALSE test. If the test is true, then one process is performed. If it is false, another (different) process is performed. In either case, there is a common single exit.

All three of the basic blocks have a single entry and exit. In fact, most of the derived benefits relate to this very fact. When first introduced, structured programming was often referred to as GO-TO-less programming. This is in reference to the GO-TO statement of FORTRAN and the penchant of programmers to abuse its use. In the minds of some, the use of a GO-TO statement is the villain that ruined many a program. In reality, the problems arose from a difficulty in following the logic of unstructured programs. The lack of organization produced unreadable programs (sometimes even by the author), consequently decreased programmer productivity, and made testing more difficult. Contrary to the beliefs of many, well-structured programs can be written in a language that is not a structured programming language. Thus FORTRAN, COBOL, and other languages of the past may be the vehicles for the production of well-structured code. More importantly, poor code (unreadable due to its complexity or cuteness) can be produced in a structured programming language.

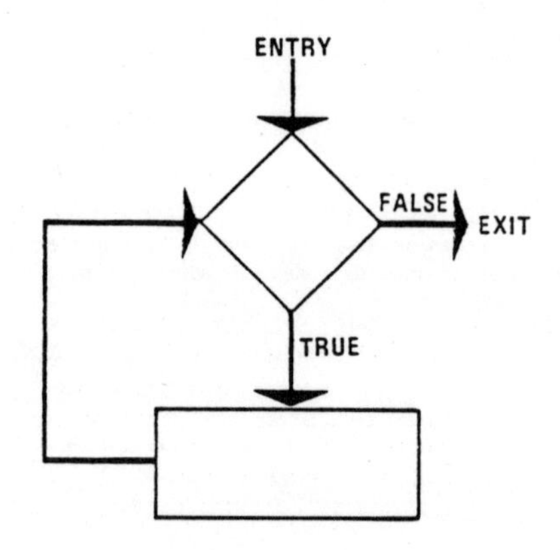

Figure 17
GENERALIZED LOOP MECHANISM

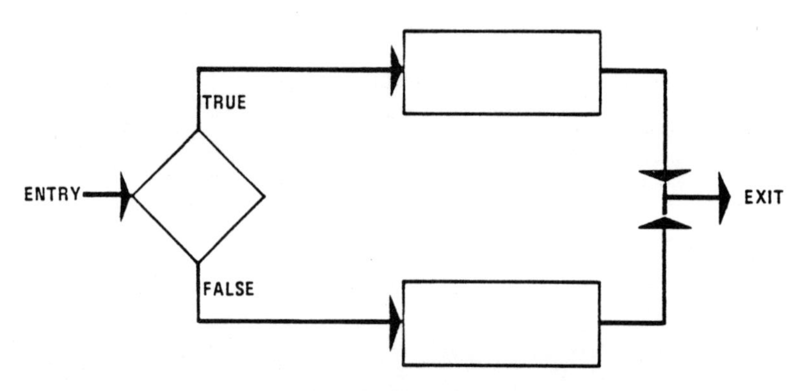

Figure 18
BINARY DECISION MECHANISM

If one examines the process box concept as first introduced, it becomes obvious that both the DO WHILE and IF-THEN-ELSE building blocks can be considered as single-process boxes (see Figures 19 and 20). This is due to the single entry and exit rule imposed. Thus, nesting of the IF and DO capabilities is both possible and encouraged.

Using the proper building blocks is necessary but not sufficient for the production of well-structured programs. One must pay close attention to the goal of producing readable code. The following are guidlines to that end.

1. Program modules must be small. A good general rule is that the whole module should fit on one page. When this is not possible, in almost every case the module being coded can be functionally subdivided to make it more consistent with the rules of top-down design.

2. Include comments in the program. This is an area often neglected in the past. Comments should include what is being done, why, and what assumptions if any have been made. Both the comments and the program must be revised as corrections are made. There is nothing worse than an incorrect comment when a new programmer tries to make a revision. Comments should not be the obvious, but should be only those comments that are helpful to a person trying to read, modify, or otherwise understand the program. Excessive or trite comments often obscure the value of those that might help a reader.

3. Don't misuse the instruction set or the software language. The programmer who takes advantages of oddities or other little known and rarely used aspects of the machine or language tends to obscure the meaning of the program and reduce its portability. Furthermore, if the oddity or undocumented feature is ever changed in future releases of the hardware or software, a very hard-to-find error suddenly appears.

4. Don't write programs that modify themselves as they execute. This rule has many motivations: to preserve clarity, to allow for reentrant code, to permit simultaneous execution from a commom area, and finally to allow for simpler testing. (This means to avoid the use of the ALTER statement in COBOL, the assigned GO TO in FORTRAN, and mixing of variables and labels in PL1, etc).

5. Avoid complex arithmetic statements. Use of complex operator precedence in writing programs opposes the structured programming discipline. Always use parentheses and break up long assignment statements into several steps. Even though the compiler and computer will get it right, more often than not the reader will have an easier time with simpler statements.

6. Format the program so that listings are more readable. Indent and use several lines for IF-THEN-ELSE and DO-WHILE constructs. The time it takes a programmer to format the code for clear reading is often more than compensated for when he has to test and debug the program. Thus the savings to readers and modifiers of the program are a bonus.

7. Try to avoid negative Boolean logic. Frequently the addition of a NOT in front of an expression is confusing to a reader. In general, a reversal of the THEN and ELSE clauses permits that the NOT be dropped. For clarity, avoid NOT when possible.

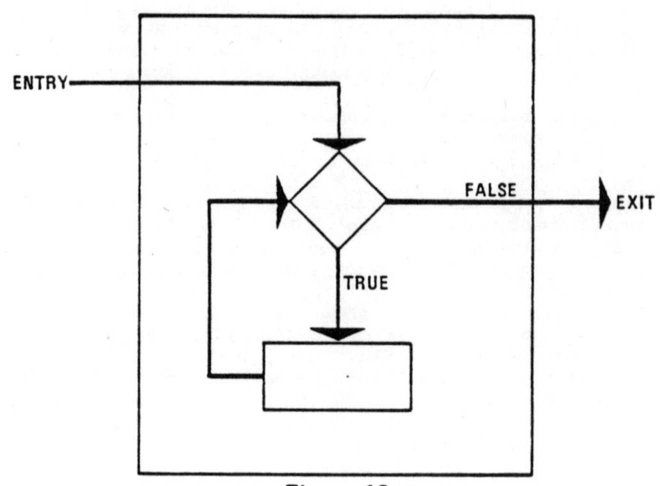

Figure 19
"DO WHILE" EMBEDDED IN A PROCESS BOX

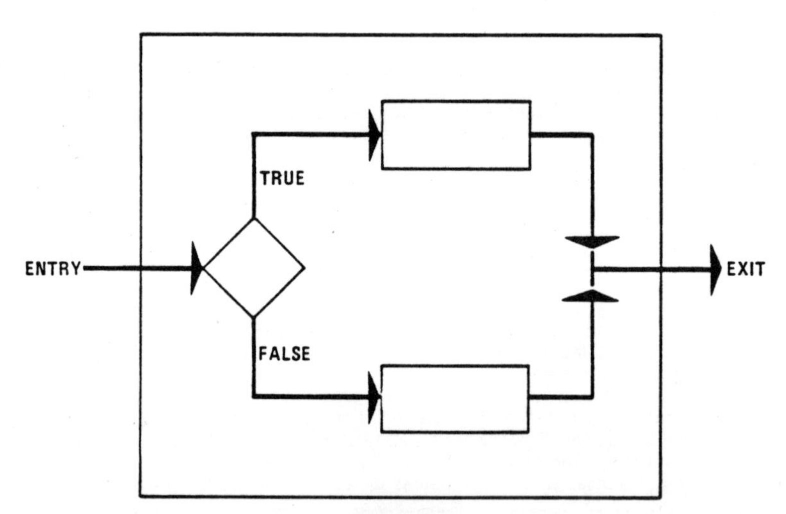

Figure 20
"IF THEN ELSE" EMBEDDED IN A PROCESS BOX

8. Use meaningful names for variables and procedures. The aid of a variable whose name has a clear meaning is invaluable in trying to read and understand a program. Most of the newer language implementations allow more than the cryptic six- or seven-character names of the past. Even on older compilers, there is no excuse for the single character names found in many programs.

9. Never allow one module to interfere with the code of another. Avoid the use of shared variables and implicit connection between modules. The goal here is to preserve the modularity of the code. It is sometimes necessary to have more than one program access a data base, and common storage can facilitate that. Extreme care and much documentation should accompany such implementations. Where possible, access to common data items should be made by the use of procedures that maintain the data. In no case should programs store local values in common areas; this has been the source of many a hard-to-find bug.

In conclusion, the key to producing quality software lies in the discipline which is exercised in its creation. The important concepts are the top-down approach to produce a modular design, followed by an implementation by the use of structured programming techniques. The clarity of the program lies in the readability of both the design and program modules. This is particularly important for maintainability, especially if those who maintain the software neither designed nor wrote it. Given that programs can be read and understood, it follows that they can be modified and the implication of the modification will also be clear.

5.1.4 Program Development

For a complex robotics program to be developed successfully, a development methodology must be established. It should be made easy for the programmer to review what he has told the robot to do at whatever level of detail he finds convenient. A listing of the current version of the program should be available and it should be easy for the task programmer to get his program into the control computer. Assisting the task programmer in every way in debugging his program is equally important (Reference 74). This means making it easy for him to try out his program without undue risk to the equipment. The same interactive debugging techniques that have been invented to aid in the development of conventional computer programs are also useful in debugging robot control programs. These techniques include

- Close control over the program as it runs, including the ability to run the program slowly or one step at a time (single stepping) and to change the current point of execution

- Ability to display and modify the current values of data items in the program, preferably by name, and, if possible, allow them to be set to the value of an arithmetic or logical expression

- Ability to specify locations at which to stop normal execution and give control to the programmer for debugging. Such a location is called a breakpoint

- Ability to record information about program execution during normal operation (logging or tracing options) -- for example, storing the line number of each statement executed in a memory buffer in order to trace the actual flow of control through a complex program.

In some industries, a workpiece represents a considerable investment. Minimizing the number of workpieces consumed in debugging the robot's task program is important. This leads to two requirements: minimize the total number of bugs in the program to begin with and find as many of them as possible per workpiece turned into scrap.

Simulations and interactive graphics can help discover errors in a program without actually operating the equipment in the robot's work station (References 75-78). They also allow the robot to continue producing the previous batch of parts while its next program is being debugged. Such program development facilities are, of course, very expensive to develop and require considerable skill to use. A simulation can never include all the details of the actual situation in the factory (until factories are redesigned) so the first trial with the real equipment will still be risky.

Many computer programs to simplify the process of producing animated films have been written in research laboratories and universities. These programs usually support a specialized programming language for describing images of objects and motions that they are to make in the film. Some of the techniques used in these animation languages may well be of use in robotics as an aid in producing animated, real-time displays of what the work station would do if it were given a specific task program to carry out. GRASS and SAMMIE are two languages specifically adapted to graphic simulation of robots in a work station in terms of their shapes and motions (References 79,80). Higginbotham is using SAMMIE to evaluate the suitability of various robots for specific tasks via computer graphic simulation (References 75-78).

Some advanced language translation techniques such as strong typing can help in discovering certain common kinds of semantic errors when the task programmer first enters his program. However, these methods are usually only used with formal programming languages such as PASCAL, RTL-2, MODULA, LIS, and the forthcoming Department of Defense language, ADA (References 60,61).

One approach to finding as many bugs as possible for each workpiece used up is to arrange matters so that during a test run, the task programmer can correct the algorithm of his task program without having to start his program from the beginning (which would probably require him to get a fresh workpiece). This process has been called hot editing, and it permits the robot to continue working from where it left off on the same workpiece after the task programmer fixes a bug. This is not too difficult to achieve if the task program has a relatively shallow procedure call depth. The task programmer can then skip over the early parts of the program with relative ease in order to get back to where he was when he hit the last bug. However, in complex programs, it may be necessary to "keep one's place" in a deep nest of calls while modifying the code describing the task algorithm. This can be difficult to manage since the code modifications must not invalidate any current bindings between formal parameters and actual parameters in those calls, and all the return addresses for the calls must remain valid. Luckily, such sophistication will not be needed in the great majority of task programs that will be written in the next couple of years.

Two fundamentally different modes of programming may be described as on-line and off-line programming. On-line programming ties up the actual production equipment during program development while off-line programming allows the production equipment to continue performing a productive task while its next task is being programmed.

On-line programming involves using the actual production equipment to demonstrate procedures or define values of data items in a task program. On-line programming necessarily takes the tooling out of useful production and therefore incurs the double expense of lost production and (generally) wasted workpieces. Contemporary record-playback methods of programming robot arms are good examples of this mode of programming.

Off-line programming involves activities such as writing task programs in some programming language, running simulation programs to test task programs without the risk of damage to the real equipment, and the collecting and organizing of large data bases to be used by other programs in generating task programs. Much of this programming activity can be performed by an engineer or robot trainer at a desk, and time-sharing techniques can allow many such people to develop work-station task programs simultaneously at a relatively low cost per person.

On the basis of the level of detail that the programmer must put into the task program, a distinction can be made between explicit and implicit programming. Explicit programming is the normal mode of programming a robot manipulator or NC machine tool. In this mode, a person specifies in detail each and every action that the machine should perform.

Implicit programming, on the other hand, is very much in the research stage at the present time. In implicit programming, the programmer would describe the work station's task in a much more general, high-level, abstract way than he would in explicit programming.

What makes implicit programming methods even thinkable today is the degree of success that has already been obtained in various artificial intelligence research efforts. In these preliminary studies, the fact that implicit and explicit programming techniques require (very roughly speaking) the same amount of information in order to generate successful programs of a given level of complexity has already become apparent. However, the amount of information tends to appear as data in implicit programming and as algorithms in explicit programming. This result is actually encouraging because the aerospace industry is one of the leaders in the development of computer-aided design methods (CAD). CAD systems seem to be the tool to create and manage the data bases that will be needed in the future for implicit programming.

5.2 CONTROL FUNCTIONS

A major part of a robotic control system is the functional element structure that provides the information handling capability for decision making.

This section presents a review of some of the concepts, methods, and practices that can be used to accomplish various control functions. Topics of discussion are work-station decisions, tooling status information, mass data storage, and external tool control.

5.2.1 Work-Station Decisions

The more decisions the work station can make at run time, the better it can adapt to changing circumstances in order to maintain a high production rate. One (idealistic) extreme is the full NC approach in which a supervisory computer plans out every detail of the work station's job, and the work-station computer has no decisions to make at all.

An automatically programmed tool (APT) programming system can then, in principle, generate explicit instructions for every manipulator motion and NC tool action, and no sensors would be needed in the work station. Actually, several conditions in real factory situations will make it advantageous to delegate to the work station responsibility for some details of how to accomplish a manufacturing task.

There are four kinds of decisions the work-station controller may have to consider at run time. These four kinds of decisions are how to determine values for deferred data items in its task program, how to allocate resources, how to coordinate concurrent processes, and how to react to exceptional situations.

5.2.1.1 Deferred Data Values

Data items whose values are determined only at run time are called deferred data items. These items allow the work-center computer or human programmer to plan work-station manufacturing procedures without having to know exactly where every object will be in the workspace. Deferred values are determined by training, sensing, or by computing them based on the values of other data (that may in turn be deferred). Training the value of a deferred data item requires interaction between the equipment and a person. The person might, for example, move the manipulator manually to a location in the workspace, type in part numbers on a keyboard, or select processing options from a menu on a display screen. Some ways in which the work station might use sensors in order to obtain values for deferred data items include: locating an object visually with a television camera, feeling for the location of an index mark or a jig (e.g., a ball or corner), and using an optical character reader to read information that has been silk-screened onto a workpiece.

In the full NC approach, the position of every object in the workspace must be accurately known before the work-center control computer can generate the manufacturing procedure that the work station is to carry out. Furthermore, nothing in the workspace can be moved unless and until that procedure specifically calls for it to move. In practice, equipment movement during maintenance, a forklift truck collision with equipment, or the various shapes of the workpieces may make this impractical. Then it becomes advantageous for the work station to determine the precise position of everything at run time.

If the values of some data items are to be computed from the values of others at run time, then planning is required to ensure that all of the data values to be used in the computation will be known when that computation is performed. Since allowing the equipment operator to determine the sequence of various set-up or training activities may be advantageous, preventing him from causing a deferred data item to be used before its value has been determined may be impossible. To guard against this possibility, the work-station controller should maintain at least one bit of status information for each deferred data item in order to indicate that the item's value has not been determined.

5.2.1.2 Resource Allocation

Resource allocation problems arise whenever two or more processes require the same resource and whenever one process can use any one of a group of different resources for a given purpose. A work station may have several pieces of similar equipment, such as vision subsystems, drilling tools, or buffer storage areas for workpieces. At run time, some of these items might be out of service or assigned to other ongoing tasks. The station's productivity would be increased if the controller could simply select another vision subsystem, drill, or storage area and proceed with the job.

Some resources that a work station will probably have to allocate include arms, cameras, end effectors, jigs and templates, space in the work area, and the attention of the work-station control computer.

5.2.1.3 Coordinating Concurrent Processes

Concurrent processing refers to a mode of operation in which a computer simultaneously executes code from more than one location in a body of code. The work-station control computer may use this mode when operating two or more pieces of equipment at the same time is advantageous. Two examples of a two-handed coordination task are placing a sheet of metal into a brake and positioning a fuselage panel assembly in a riveting machine. Concurrent processing may be done either by multiprogramming (running several programs in one computer), or by multiprocessing (several computers running different parts of the same program).

Sometimes two or more activities can take place simultaneously because they can share one or more scarce resources. The point in the program where the computer begins to perform those activities is called a fork. Sometimes an activity cannot be started until two or more other activities have been completed. The point in the program where the flow of control comes together again is called a join.

The coding techniques used to coordinate concurrent processes are conceptually quite simple. First of all, a multiprogramming mode of operation requires some method for deciding which activity the computer's central processing unit (CPU) will work on at each instant. The code which switches the attention of the CPU is called a scheduler, and the routine does not need to be very complicated. One way of dividing the attention of the CPU is simply to allow it to work on any activity until it has to wait for some external event to occur, and then switch its attention to another activity that is not waiting for such an event. A second way is to allow the CPU to work on any activity for only a certain amount of time (usually a fraction of a second), and then switch it to another activity. Some activities that are more vital to success can be assigned higher priorities than others. Careful planning to ensure that the work-station computer will be able to give enough service to the set of concurrent tasks is required to ensure that critical ones are never prevented from running.

Often two activities will need the same resource (perhaps a vise). An efficient way to share the resource between those two activities (and others) is to include in the work-center control program a special data item associated with that resource, a semaphore or Dijkstra flag. A semaphore is just an item of status information that indicates whether or not the vise is in use. Semaphores should not be accessed like ordinary data items, however, because in order to acquire the use of the vise, an activity has to test the semaphore's value to see if the vise is free, and if it is free, set the semaphore's value to show that the vise is now in use. The work-station control software must be written so that no other activity can try to do anything with that semaphore while the first activity is testing and possibly setting it. For example, if the scheduler in the work station switches the attention of the CPU from activity to activity on every clock interrupt, then it might do so between steps one and two. The second activity might, by a stroke of extreme bad luck, attempt to acquire the same vise. Noting that the vise was still free, the second activity might set the semaphore to read IN USE and then begin using it. When the first activity got a chance to run again, it would not know that the semaphore's value had been changed and would proceed to set the semaphore a second time and then try to use the vise, also. This situation could cause serious damage to the equipment and workpieces.

One way to prevent such errors is to lock out clock interrupts for a few microseconds while the first activity both tests and sets the semaphore. After that, the clock interrupt can be released so that the next activity can test that semaphore. The semaphore will show the vise to be in use by the first activity, and the second activity will then know that it must wait for that tool to become free again.

Such interrupt lockouts are unnecessary in computers that can, in a single instruction, both test a semaphore and conditionally set it. Most computers service interrupts only between executions of individual instructions and effectively lock them out automatically during each instruction.

One of the major problems in parallel processing is called the deadlock, the deadly embrace, the circular wait, or the interlock. The use of semaphores will not prevent deadlocks. Deadlocks occur whenever two activities that are proceeding concurrently each tie up a resource that the other needs. For example, the two activities might be

1. Use a camera to locate the edge of a stringer, pick up a drill with the gripper, and drill a rivet hole a certain distance away from that edge.

2. Use the gripper to pick up the camera and search for a tooling mark on another workpiece.

The resources in this example are the camera and the gripper. Tasks one and two might both get started by a fork in the main program. Task one might request and be granted the use of the camera, after which task two might request and be granted the use of the gripper. At that point, the deadlock has occurred. Task one needs the gripper, but can't get it from task two. Similarly, task two needs the camera that task one has. Consequently, the work station will stop working.

One way to prevent such a deadlock would be to detect the possibility of it at the time the work-station computer's instructions are being planned, before that computer even receives them. Either the work-center computer or a programmer can check for the possibility of a deadlock. If one should be found, the plan can then be revised to prevent this from happening. One revision that would avoid the problem in the example above would be to do task one before task two, instead of attempting to run them concurrently. Another solution would be to revise task one so that it would relinquish the use of the gripper when it could not obtain the camera and bid for them again every few minutes until task two finished with them both.

Some ad hoc methods have been developed to prevent deadlock at run time. They are not applicable in all situations, but they might be of some use to the work-station control computer. One method is to require all activities to request their resources in a certain order. Another method is to set a maximum limit on the length of time that an activity can retain any resource.

The deadlock problems described above are contention deadlocks because they arise from activities contending for scarce resources. Other kinds of deadlock can occur as a result of timing relationships in programs. Ensuring that these timing deadlocks can never occur in a given system that has critical timing constraints is very difficult. One way to reduce the probability of their happening is to divide the software into modules that interact totally asynchronously. Timing deadlocks between events in different modules will then be improbable, and any timing deadlocks that do occur can be easily located to events within one single module.

5.2.1.4 Exception Handling

Exceptional situations are conditions that require a response from the work-station controller. These situations fall into two categories: predictable and unpredictable. Some predictable situations are

- Slow loss of manipulator positioning accuracy (drift)
- Worn cutting tool
- Incorrect or defective workpiece
- Workpiece not in correct position
- Defective fastener
- Next workpiece not available
- No place to send finished workpiece.

The work-station control computer will only have to test for conditions like these infrequently -- at a few stages during each production cycle or a few times during each shift.

Unpredictable exception conditions typically can occur only during well-defined intervals during a manufacturing task, but during that time, they can occur at any moment. Furthermore, when these conditions do occur, they present such an immediate danger of a mishap that the work-station controller must respond to them instantly. That computer must very frequently monitor for each such condition during the entire time that the condition may possibly occur. Some examples of unpredictable exception conditions are

- Intrusion of a person or an object into a hazardous area
- Collision between the manipulator and something else
- Sensor failure
- Cutting tool breakage
- Workpiece breakage
- Power outage
- Pneumatic or hydraulic line rupture
- Object fallen out of gripper.

Where possible, the use of special-purpose hardware to monitor continuously for unpredictable conditions and cause an interrupt signal to the work-station controller the instant that they occur will be advantageous. The alternative is to complicate the design of the software in the work-station control computer -- for example, by requiring the computer to set up concurrent processes that will read and interpret dozens of hazard-sensor signals and scores of times per second while operating the automatic equipment.

Exception-handling code should require the fewest possible resources. Any resource required may be in use when the exception condition occurs. Even if that resource can be freed temporarily for use in correcting the exception condition, freeing it will take time. The unpredictable exceptions in particular will usually require an instantaneous response, so there will be no time to obtain any resources.

5.2.1.5 Automatic Indexing

Automatic indexing is the process of determining the position and orientation of a workpiece with respect to an automatic machine, such as a manipulator, that is to perform some operation upon that workpiece. By assumption, the machine's control system can use the locatic n information in carrying out those operations. The control system might use the measured displacement and rotation to modify each preplanned action, such as a manipulator motion during run time, every time it performs that action. Alternatively, the system might make a single pass over the descriptions of all such actions, modifying each one time only before beginning the production run.

Aerospace manufacturers will probably be wise to agree (at least within their own plants) upon a standard method for indexing (i.e., measuring the position of a jig with respect to a manipulator). The method chosen should have at least the following characteristics:

- The procedure should be entirely automatic, and capable of locating a jig under control of the work-station control computer. This eliminates the possibility of human error and will result in more uniform performance.

- The procedure should require only simple and inexpensive tooling components on the jig itself because a set of these components will have to be permanently mounted on each jig.

Measuring the positions of three noncollinear points is sufficient to measure the position of a rigid object in space. Thus, a jig could have fiduciary objects, whose locations the work-station control computer could measure accurately with a sensor welded to it at each of three widely separated places. For accuracy, the fiduciary objects should be as widely separated as possible yet still be within reach of the manipulator.

A simple method (illustrated in Figure 21) that could be used is discussed in the following procedure. Weld a steel rod about three inches (10 cm) long and about .5-inch (1 cm) in diameter to each of three points on the jig frame. On the free end of each rod, attach a one-inch diameter steel ball. The rods should be approximately parallel and should point toward the manipulator side of the jig. The centers of the balls define the three points whose locations are to be measured for indexing. Adopt a convention for the order in which to measure their positions (e.g., clockwise from the upper left ball). Use an end effector that carries a corner probe to measure the position of each ball. This probe is an inside corner that faces away from the wrist socket. This corner could be milled from a block of steel or fabricated by welding together along their short edges three right isosceles triangles of sheet steel. Assuming that the positions of the balls are known approximately, bring the corner probe into contact with each one individually as follows: Hold the corner with its faces normal to the axes of the Cartesian coordinate system in which the manipulator moves and with the interior of the corner facing away from the manipulator toward the jig. Position the corner about one inch (3 cm) to the right of the first ball with the interior of the corner facing the ball.

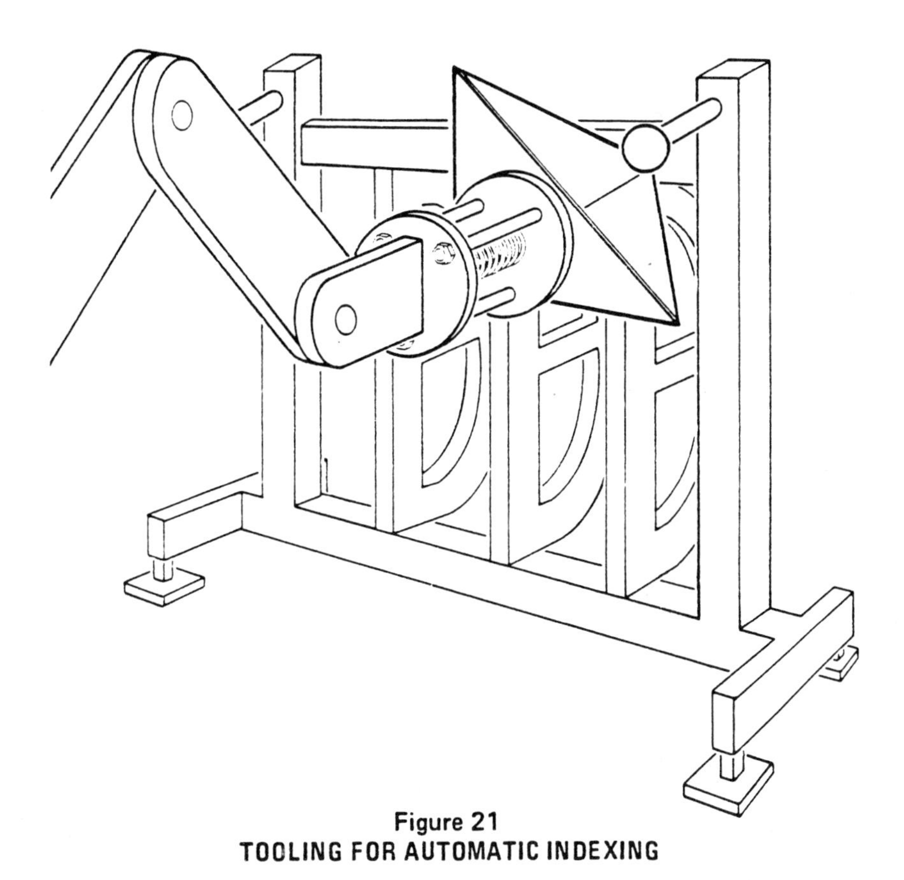

Figure 21
TOOLING FOR AUTOMATIC INDEXING

Move the corner slowly to the left until it touches the ball. (Contact can be sensed electrically by completing a low-voltage circuit through the manipulator, corner, ball, rod, and jig frame.) The position of the center of the ball is then .5-inch (1 cm) to the left of the vertical face of the corner.

The above procedure provides one horizontal component of the ball position. Repeat the procedure by approaching the ball from above in order to determine its height. Finally, advance the corner toward the ball until it makes contact in order to determine the remaining horizontal component of the ball's position. Repeat this procedure for each ball.

The corner-shaped sensor should be supported by a breakaway mounting of the preloaded-spring type described in Subsection 4.2.3.1.3. This support will prevent damage if the manipulator should fail to stop soon enough after contact and will return the corner accurately to its normal position when the manipulator backs off.

The balls do not need to be welded to the jig in precise positions, and they do not need to be located accurately on vertical or horizontal lines. As long as they are not collinear, the three ball locations uniquely define a Cartesian reference frame (called the jig-reference frame) in a fixed position with respect to the jig according to the following algorithm:

Let the three-component Cartesian positions of the balls in the manipulator's coordinate system be the three vectors, B1, B2, and B3. Let JX, JY, and JZ be the three orthogonal axes of the jig-reference frame to be defined (Figure 22). Then, the following apply:

- The origin of the jig axes is at B1.
- JX is the axis from B1 towards B2.
- JY is the axis through B1 at right angles to JX and in the plane that contains JX and B3.
- JZ is the cross product of JX and JY.

In general, this will result in a jig-reference frame that is displaced and rotated, perhaps considerably, from the reference frame in which the manipulator operates. This is perfectly all right since the robot's computer can easily convert a position given in the jig frame into its own frame of reference, and then to joint angles.

Once these simple computations have been made by the work-station control computer, training of new positions or playback of previously trained positions may proceed. The work-station controller should record all positions in terms of their JX-JY-JZ coordinates and should convert between them and the manipulator's normal X-Y-Z coordinates as needed.

The method described above for locating the fiduciary objects (the balls) assumed that the work station knew their approximate locations before beginning the search. This information could be obtained in a variety of ways.

The workpiece tooling may simply constrain the jig to always be positioned to within an inch or so of a standard position. Then, knowing which jig it is indexing from, the work station can look up in its data base for the approximate locations of the fiduciary objects on that jig. Alternatively, the work station could actually locate them by searching with an appropriate sensor.

A camera might be an appropriate sensor if an easily identified visual target pattern (e.g., a bull's-eye or a Maltese cross) is painted on the jig near or around each object. A camera situated at (or carried by the manipulator to) a position where it can see the entire jig could quickly locate the patterns to within at least a few inches. A second picture taken close up might be necessary in order to refine the estimate before beginning the tactile search.

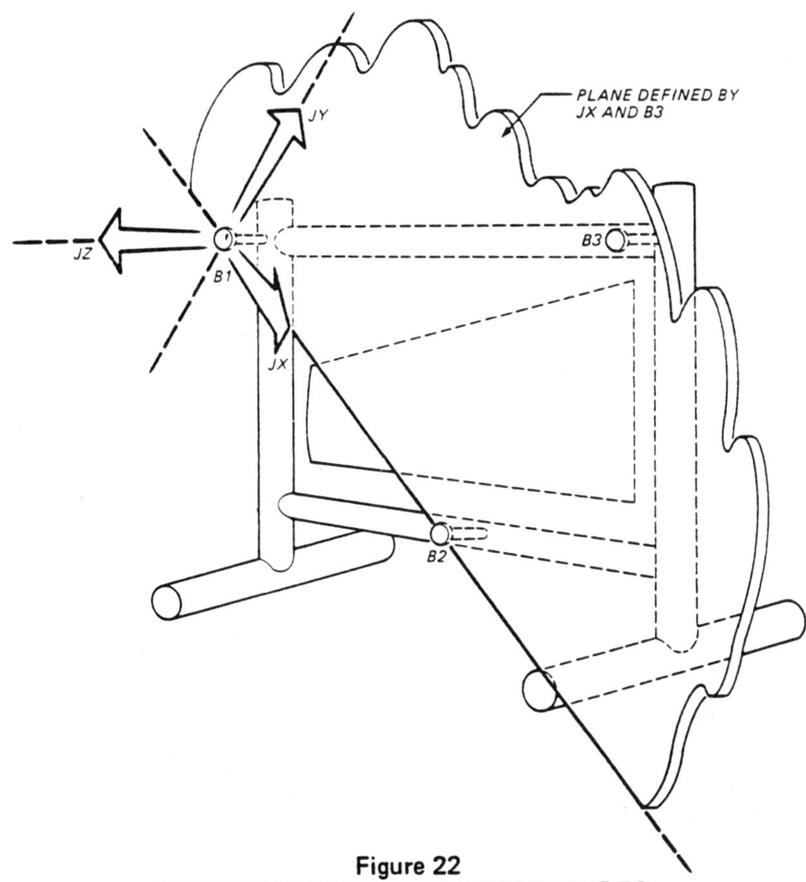

Figure 22
CONVERSION OF JIG REFERENCE FRAME TO CARTESIAN REFERENCE FRAME

Another interesting type of sensor would be a magnetic-field sensor which would react to the field from a permanent magnet in each ball (Figure 23). A field-direction sensor would be more expensive than a simple field-strength sensor but would allow the manipulator to find the magnet by traveling along the field direction. That method would be quicker than finding it with a hill-climbing search method.

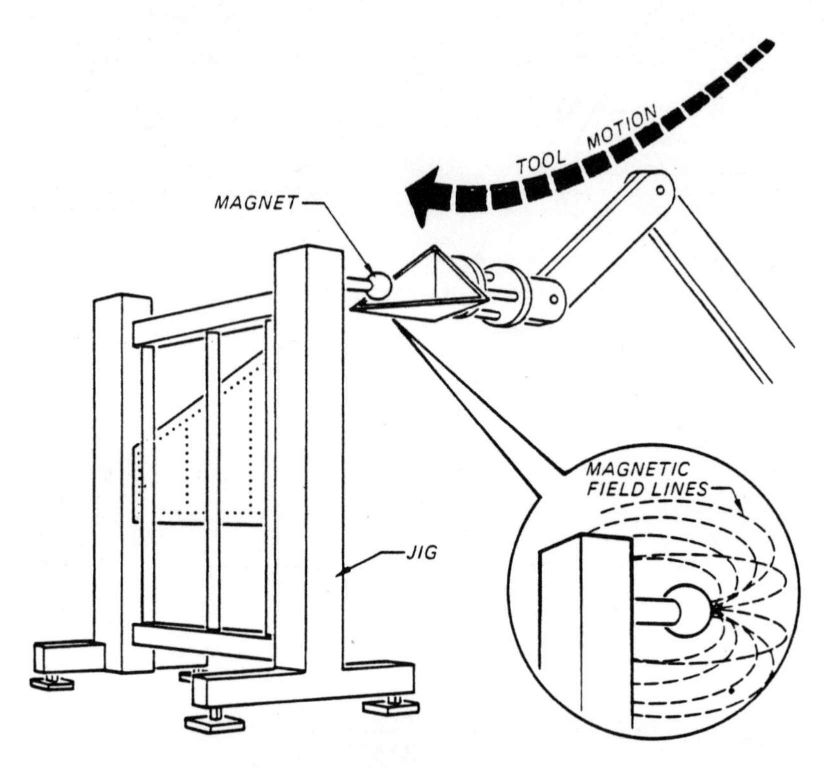

Figure 23
LOCATING A FIDUCIARY OBJECT IN A JIG BY FOLLOWING A MAGNETIC FIELD LINE

Yet another method would be to mount light-emitting diodes near the fiduciary objects and search for them with a photocell in the end effector. In order to eliminate interference from natural sources of light, standard practice in such applications is to filter out all light except the color emitted and to amplitude-modulate the light emitted and detect it synchronously (Reference 25). Commercial infrared sources and detectors that do all these things are available. In order to locate the emitter quickly, a cylindrical lens can be placed in front of the photocell so that it sees light only in a plane normal to the axis of the lens. Two sweeps of this plane -- one horizontally and one vertically -- should suffice to locate the emitter to somewhere on a line, and a third can then determine its position along that line by triangulation. To lower jig costs, reflector targets could replace the emitters on each jig, and one emitter could be located in the end effector.

5.2.2 Tooling Status Information

In order to determine the next action, the work station has to know what courses of action are open. This requires knowledge of what is currently going on in its work area, what resources are currently in use, and what resources are free. The work-station control computer should maintain up-to-date records of the status of all tooling in its domain.

The more responsibility the work station has for deciding what it should do from moment to moment, the more of this status information it will need to keep in storage in order to be able to make those decisions. Different classes of tooling require different amounts and kinds of status information. The following sections give examples of status information that would be useful in controlling specific tools.

5.2.2.1 Fixed Passive Tooling

In order to control a fixed passive tool, the work station has to remember where the tool is and how it is being used. The amount of information that must be stored in order to describe the tool's state will depend upon the tool and how it can be used by the work station. For example, a simple metal jig that can hold one workpiece of a specific kind has only two possible states -- holding such a workpiece or not holding one. These two states can be completely represented by just a single bit in the work-station software. On the other hand, a large worktable might have room to hold several objects. For some tasks, the work-station controller may need to remember what objects are currently on that table and where they are located. That would require a much more complex data structure in the work-station software -- perhaps a one-dimensional array of records, each describing one object on the table and giving its location.

5.2.2.2 Fixed Active Tooling

A fixed active tool also requires that the work-station controller remember where the tool is and what it is currently doing. In addition, remembering what control signals were last sent to that device are sometimes useful. For example, a pneumatically operated vise could be controlled by a single binary OPEN-CLOSE signal from the work-station controller. Then it might be important to distinguish between the following four states of the vise: open-empty, open-occupied, closed-empty, and closed-occupied. Two bits of state information would suffice to represent these different states. Since a vise can hold many different kinds of objects, additional information about the object (if any) that the vise is holding would probably be useful.

5.2.2.3 Movable Tooling

Movable tools are, in general, more difficult to control than fixed ones. One problem is that in order to use a movable tool, the work station must allocate a manipulator to carry it. The work station may, for example, have to decide which manipulator to use for carrying the tool, or it might have to decide which tool to pick up with a given manipulator.

Another problem in controlling movable tools arises whenever the manipulator sets such a tool down on another movable object. For example, the manipulator might temporarily set a workpiece, tool, or jig down on a tote box, conveyor, or part positioner. If the work-station controller later causes the supporting object to move, the supported object will move with it. The next time the supported one is needed, the controller will have to determine where that object went. Similar control problems arise whenever any two movable objects become rigidly attached to one another, such as when a template is pinned to a workpiece. To deal with this kind of control problem, the work-station controller must remember which objects are attached to which and perhaps even how they are attached.

A tree structure is a convenient data format in which to represent attachment relationships (shown in Figure 24). Each node in the tree can represent one object, and each branch can represent an attachment relationship between two objects. The branch would describe at least the relative positions of the objects and perhaps some indication of how they are attached — i.e., rigidly, by gravity only, or by a permanent or temporary fastener. In order to find an object, the work station would search down the tree from the node representing that object until it found a node for some object whose position was unknown. (At worst, it would reach the root node that represents a point at a known location on the work area floor.) The sequence of branches followed would then imply a chain of relative positions from which the work-station controller could compute the current location of the object that it was seeking.

5.2.3 Mass Data Storage

The work-station control computer will need mass storage for at least two and possibly three types of information: task program and data, run-time data, and system software. Four classes of mass storage devices are practical for aerospace use in storing this data in the work-station computer system.

5.2.3.1 Work Station Data

The task program and data and the run-time data constitute the software that is specific to the manufacturing task that the work station is to perform, such as drilling rivet holes in a wing panel. The task program is an algorithm for performing the task - a description of the events to occur during processing and when they should occur. This description will generally include repeated sequences (loops) and alternate sequences (branches). The task data and the training data are the quantitative information needed to perform the program. This includes information such as the location of the wing panel, where to drill holes in it, and how many holes to drill before changing the drill bit.

In the Integrated Computer-Aided Manufacturing (ICAM) model of a hierarchical computer system for production control, the task program will be generated by the work-cell computer (Refer to Subsection 5.3.5 for an explanation of the ICAM hierarchy). Some of the quantitative values needed can be supplied to the work station along with

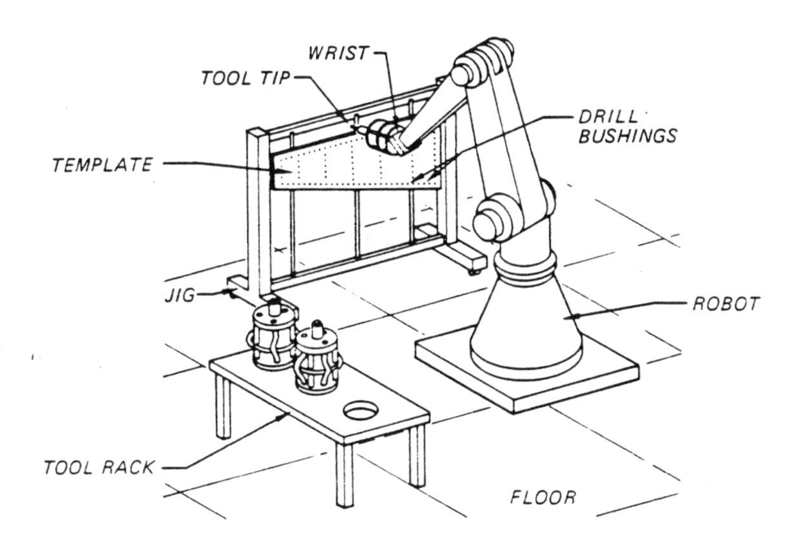

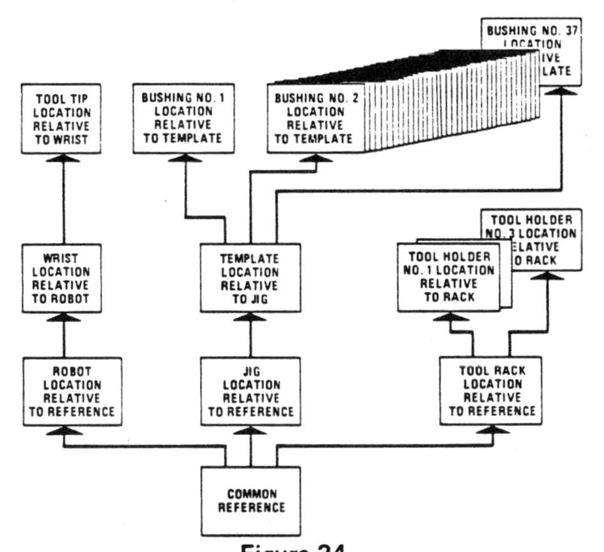

Figure 24
TREE STRUCTURE REPRESENTATION OF ATTACHMENTS, RELATIVE POSITIONS, AND ABSOLUTE POSITIONS

the algorithm, but some will only be determinable after the work-station computer receives its task program. These are the deferred data values, as discussed in Subsection 5.2.1.1, which deal with the placement of equipment in the station, the calibration factors of the various sensors, and the shapes and sizes of the end effectors, etc.

The deferred data values may be further classified into training data and run-time data. Run-time data is that read from a sensor during normal operations, usually in order to make a decision about which of two alternate sequences of processing steps to perform. Run-time data items may be assigned new values repeatedly during a production run. On the other hand, training data is put in once during start-up of the work station, usually with the help of a human operator or trainer.

Run-time data might consist of information about equipment or workpiece placement, the appearance of objects to be recognized or inspected later, or weight, force, and torque limits to be observed during operations. This sort of data can be typed in as numerical values by the operator, but this is usually undesirable for the following reasons: obtaining the numerical values may require time-consuming setup and operation of measuring equipment, and the process of typing in the numbers is slow and prone to error.

Using the sensors in the work station is a much better way to measure the values of as many of the training data items as possible. The work station may even have enough prior information to verify whether or not the values being read are reasonable. With properly designed software in the work-station computer, the process of obtaining the training data can be a rapid process in which the computer supplies much of the expertise required, and the human trainer supplies relatively little. Specifically, since the work station knows what items of information must be trained, it can tell the person what to do to supply it.

5.2.3.2 Mass-Storage Devices

Four different kinds of mass-storage devices that will be practical for use in aerospace work-station computer systems include: magnetic tape, magnetic disc, bubble memories, and random-access memories. Of these types, bubble memories will probably prove to be the most desirable in the long run. At the moment, conditions in the peripheral-device market have still not permitted sufficient emphasis on the development of magnetic-bubble technology by the semiconductor industry. Only a few bubble memories have even been marketed as yet.

The major considerations in choosing a mass-storage device will be

- Size - The amount of information that it can hold
- Speed - The rate at which the information can be read from or written to the device
- Access Time - The time required for the computer to read or write a unit of information from an arbitrary location
- Cost-Per-Bit Mounted - The cost per unit of information stored and immediately accessible to the computer
- Cost-Per-Bit-Dismounted - The cost per unit of information stored but not immediately accessible to the computer

- o Reliability - The mean time between failures in the manufacturing environment
- o Hardware Support - The cost and availability of repair and/or replacement units; service ratio, diagnostic softwareoSoftware Support - The operating systems, file management routines, device drivers, etc., that make the hardware useful.

To a first approximation, the four storage technologies mentioned above occupy the same relative positions along a spectrum of capabilities in the areas of size, speed, and cost. The size decreases, the speed decreases, and the cost-per-bit increases in the following order: magnetic tape, magnetic disc, bubble memories, and random-access memories. Although the range of capabilities within any one technology is relatively large, there is little overlap between adjacent members of the list in terms of their capabilities; the four devices have distinctly different characteristics, and consequently are usually used for different purposes.

A major consideration for an aerospace manufacturer is just how severe an environment the storage devices (or, in fact, any piece of computer equipment) will face in daily use. Luckily, aerospace plants are remarkably clean in comparison to foundries and paper mills, for instance. Aerospace plants are often air-conditioned as well. In these conditions, the more sensitive mass-storage devices may well operate almost as reliably in the work-station area as in a conventional computer room.

Training data is rather expensive to acquire because it takes time away from production and requires the labor of one or more people. Therefore, this information must not be lost accidentally due to errors in the control software of the work station. The information can be transmitted up the hierarchy to the work-center computer, but if that machine should be inoperative when it becomes necessary to reload the training data, production time will be lost. Storing the training data locally in the work station makes it more likely to operate continuously. Having a write-protection capability on the local storage device is then useful.

Write protection means that the contents of the storage device cannot be modified in any way, no matter what the control program in the work-station computer should do. Having the software turn on the write protection is convenient, but manual intervention should be required to allow writing to occur again. For complete protection against software errors, control of write protection must be manual because the computer might not turn it on at the proper time.

Except for random-access memory, mass storage can be obtained in portable free-standing packages (sometimes called volumes), such as tape reels and disc packs that can be manually loaded into read-write hardware in the computer.

The following sections discuss some specific characteristics of each type of storage device from an aerospace manufacturer's viewpoint.

5.2.3.2.1 Magnetic Tape - Magnetic tape comes in several different sizes of reels, cassettes, and cartridges. Small tape volumes may be able to store only a few tens of thousands of 8-bit bytes, while the highest-density 2400-foot (730 m) reels can hold about 180 million bytes. The more expensive, higher performance drives almost always take reels. Cartridges may offer some protection against error-causing dust and dirt. In general, the higher the bit density (number of bits of information per unit of distance along the tape), tape speed, and reel size of a magnetic tape drive are, the higher the cost is. High-density drives are very sensitive to dust and other contaminants and probably should not be used in harsh working environments. Tape drives which store information redundantly have significantly lower error rates for a given bit density than drives which do not, and they are therefore more reliable. High-performance tape drives offer error-correcting redundant coding techniques while less expensive drives with redundancy simply detect errors.

Some form of write protection is almost always available on any kind of magnetic tape. Plastic write rings must be inserted in reels to permit a tape drive to write on them. Cassettes and cartridges have plastic tabs or pins that prevent writing on the tape when they are removed.

Magnetic tape is a relatively slow medium to access, so it is best suited for infrequent, large-volume data transfers. Some good uses include loading the work-station computer with its control software, storing a task program or training data for reuse later, and logging wide-bandwidth sensor data during operation for postmortem analysis or performance later.

5.2.3.2.2 Magnetic Discs - Magnetic discs come in a variety of sizes, both in terms of the amount of information stored and physical size. Storage ranges anywhere from 256 thousand to 160 million 8-bit bytes. The smaller discs can be mailed in an envelope while the larger discs are the size of a hat box.

Discs are used where rapid and frequent access to fairly large amounts of data are required. A good use for a disc is to store overlaid work-station control software (in particular, the operating system that the computer's vendor supplied with the hardware). Task programs that are too large to fit in the available memory of the computer can be kept on a disc, and pieces of the task programs can be brought into memory individually as needed. Voluminous training data, such as visual images of workpieces, could be kept on a disc. An especially productive use for a disc would be to store instructions and error messages for presentation to the operator during training or during production -- getting these instructions and messages off a tape as needed would be too slow. Although some manufacturers offer drives that allow protection of half of the disc area, write-protection facilities, if provided, generally protect an entire disc at one time.

Generally speaking, the less the discs are inserted and removed from their drive, the more reliable they are. Some drives are hermetically sealed and contain a single permanently mounted disc. These drives often come with a second drive in which a second disc can be inserted for rapid copying to and from the permanently mounted disc and for doubling the size of the available storage.

Latency is the time needed for the disc's read-write heads to reach the position where information is to be read or written. Latency ranges from 50 to 500 milliseconds across the types of disc drives that might be used in an aerospace work-station computer. However, several preliminary accesses may be required to read and write bookkeeping information (the disc's directory) that keeps track of what is stored where. Discs are fast enough for most real-time control applications, especially if careful thought is given to the design of the work-station control software in order to ensure efficient use of the discs.

5.2.3.2.3 Magnetic Bubble Memories - Bubble memories currently available can store only about 64 thousand 8-bit bytes. Bubble memories, which are still a developing technology, promise to replace disc memories because they offer an increase in reliability, a decrease in power and size requirements, and a potentially large decrease in the cost-per-bit of storage. Thus, these memories promise to fill a performance gap between discs and the more expensive random-access memories.

The big advantage for an aerospace manufacturer would be that bubble memories should provide the degree of reliability that is typical of other solid-state devices. Disc drives, being electromechanical mechanisms, are less reliable because they contain moving parts that wear.

5.2.3.2.4 Random-Access Memories - These devices are at the top end of the performance range in terms of their speed, but their cost-per-bit is correspondingly high. Modern computers all contain some random-access memory, but in this discussion, the reference is to auxiliary memories that are not in the address space of the computer. Their interface electronics usually makes them appear to be an extremely fast peripheral device, such as a disc with microsecond latency times. These storage devices are usually an order or magnitude larger than the memories usually supplied with minicomputers (e.g., one million bytes compared to about 32 thousand bytes), and the devices are usually built of integrated circuits.

At the moment, the attraction of these devices for the aerospace manufacturer lies in their potentially better reliability than electromechanical storage devices used with dismountable media. The random-access memories also offer a speed advantage if a large amount of data should be needed very quickly during certain stages of production. However, a bulk memory is of little help to the work-station computer in processing that information, other than to provide room for intermediate results. In the future, however, larger memories may be used for sophisticated image-processing or real-time planning application.

5.2.4 External Tool Control

The following subsections discuss the operation of tools, the utility of sensory feedback, and the motion that the manipulator must make in order to operate certain tools properly.

5.2.4.1 Dumb Tools and Smart Tools

A dumb tool is one that does its job without using any sensory information. A dumb tool may require complex control signals from the work-station computer in order to operate, but it operates in an open-loop way.

A smart tool operates in a closed-loop way. A smart tool is usually an end effector, but it could be a piece of fixed tooling. The tool includes actuators and/or sensors together with a certain amount of logic circuitry (usually a microcomputer) that enables it to perform a more or less complicated processing operation by itself. Ideally, a smart tool should only have to be held up to the workpiece by the manipulator and turned on. This tool signals the work-station controller when it has finished and perhaps indicates whether it was able to do its job properly.

Smart point-processing tools are easier to construct than smart line-following or area-covering tools. A good example of a smart point-processing tool is a drill developed by General Dynamics and used with a Cincinnati Milacron T3 arm to drill aircraft panels. This drill contains an internal actuator that pushes a collar into a precisely located reference bushing mounted in a template. Compliant elements in the drill and a chamfer in the bushing reduce the requirements on the accuracy with which the manipulator must position the machine. Other internal actuators and sensors cause and detect (1) complete penetration of the bushing, (2) penetration of the drill bit through the workpiece, and (3) extraction of the bit. A microcomputer monitors the drill's sensors and controls its internal actuators. Although this computer is located externally to the drill itself and although it also controls other functions in the work station, it could as easily be a dedicated microcomputer built into the drill.

5.2.4.2 Smart Tools For Templateless Machining

Smart tools can eliminate the need for templates in many aerospace applications. The cost of templates and the cost of jigging up materials in them is significant enough to prompt development of methods that will permit templateless drilling, routing, and other processes. Over the whole reach of a typical commercial manipulator, the level of positioning accuracy required in the aerospace industry for these processes is difficult to obtain. Therefore, in the future, it may prove cost-effective to allow a smart tool to take over responsibility for all precise adjusting in its own position with respect to the workpiece.

A smart tool for high-accuracy templateless machining operations (such as drilling) over a small region of a large workpiece would require the following three components:

1. A compliant (or possibly detachable) mounting between it and the manipulator

2. A sensor that can measure the current tool position relative to reference elements on the workpiece or workpiece jig with whatever accuracy the task requires

3. An actuator mechanism that can adjust the tool position relative to the workpiece with high resolution and rigidity.

The actuator mechanism would require neither accuracy nor repeatability because it would be used in a closed-loop servo.

The sensing and fine-positioning portions of such a tool could also be packaged as a modular component usable with several different dumb tools, such as drills and one-sided riveters, in order to reduce the average cost per smart tool.

The position-sensing and fine-positioning components of such a tool could be built in many different ways. For example, the acoustic range sensors used in commercial input devices for computer graphics might be able to provide sufficient accuracy, repeatability, and resolution for the sensing function.

The devices measure the time required for an acoustic impulse to travel through the air from a spark gap to a strip microphone. The spark gap can be quite small physically, and the strip microphones may be several feet long. Two (or three) microphones at right angles to one another give X-Y (or X-Y-Z) coordinates directly without any need for geometric computations. The spark gap could be mounted near the tool tip, and the strip microphones could be mounted on the jig that holds the workpiece. A less expensive position sensor that measures the distance of two or three taut wires stretched from the tool tip to take-up reels mounted on the workpiece jig could be built. A third approach that would be more expensive, but potentially more accurate, would be to use commercial laser interferometric distance-measuring equipment. Many other approaches are possible (Reference 81).

The actuators in such a tool could be quite simple mechanisms, such as DC motors driving leadscrews. The important thing would be for the tool to be able to attach the free end of each actuator to either a nearby structural member of the workpiece jig, or to the surface of the workpiece itself (e.g., by small grippers or suction cups). The tool could then adjust its position relative to the workpiece on the basis of the signals from its own sensors. This would ensure accurate placement of, say, a drill bit, regardless of drift or compliance in the manipulator that supports the smart tool.

One difficulty in designing such a smart tool would be making sure that it could attach itself properly in the presence of obstructions, such as jigs, clamps, and holes in the workpiece itself. In principle, however, the work-station controller should know the approximate locations of such obstructions and be able to avoid them.

Smart tools are sometimes denigrated as mere gadgetry. On the contrary, the synergistic combination of smart tools with large manipulators offers tremendous potential for cost savings and increased productivity in aerospace manufacturing. This potential arises from that industry's need for many close-tolerance machining operations in small regions over large sheet metal parts. Conventional NC machine designs can supply the needed accuracy and rigidity throughout the entire volume surrounding the workpiece, but only at high capital cost (in the millions of dollars). The cost is high because those designs surround the workpiece with massive, precisely shaped, metal structural elements. Industrial robots represent an inversion of this design approach, in which the robot is a relatively long, thin, articulated cantilever that may even be surrounded by its workpiece. Consequently, a robot manipulator can provide the reach needed to deal with the scale of aerospace parts relatively cheaply but at the cost of necessarily reduced structural rigidity and accuracy. On the other hand, smart tools can provide the missing rigidity and accuracy but only over a small working region. Small working regions are, however, perfectly adequate for many important aerospace processes, such as drilling rivet holes in stringers. When a manipulator and smart tool are combined, they produce a system with high accuracy, long reach, and low cost.

In an aerospace manufacturing cell, it is very likely that a cost-effective way to control point-processing smart tools will be to use a manipulator with a long reach, high-lift capacity, but rather coarse positioning ability to hold a small smart tool up to the workpiece and allow it to adjust its own position before it begins to perform its function.

5.2.4.3 End-Effector Motions

Point-processing tools are generally easier to control than line-following or area-covering tools. For example, drilling a hole accurately with an end effector is a simpler control problem than routing an edge contour with equal accuracy (even using templates) because in order to control a higher dimensionality tool, the manipulator and work-station controller must be able to move the tool accurately along a path. This is a much more difficult control problem than simply placing the tool precisely at a fixed position in space. The level of difficulty can be concisely expressed in terms of the constraints upon the end-effector trajectory that the work-station control must meet. These constraints can be summarized as follows:

TYPES OF MOTION

Go to a point.
Follow a contour.
Follow a contour at a specified speed.

ARRIVAL CONDITIONS

Arrival at the point or contour at, by, or after a specified time.

TRACKING REQUIREMENTS

The point or contour may itself be moving.

In any motion, constraints are usually imposed on the orientation of the end effector, too. When operating certain area-covering end effectors, the work-station controller may also have to take account of the effects of overlapping coverage. Most importantly, the end effector should never overshoot the point or contour to which it is headed. Various research groups have demonstrated the feasibility of sophisticated manipulation in real time by using only inexpensive minicomputers or microcomputers. Running software that does this in the manipulator's local controlling computer is quite practical if it has one. The work-station control computer can then treat the manipulator and its computer as a subsystem with built-in line-following and tracking capability.

The following sections discuss the types of end-effector motions required for contour-following and template-following tasks, as well as specific motion requirements for some particular aerospace manufacturing processes.

5.2.4.3.1 Contour-Following - The contour-following types of motion are used in operating both line-following and area-covering end effectors. If the arm solution equations of the particular manipulator being used have multiple solutions or singularities for some manipulator posture, then those postures will have to be avoided when following a contour. The work-center control computer may possibly foresee these problems and plan around them. Alternatively, this motion may offer the work-station control computer several alternate trajectories for certain motions and allow it to select, on the basis of the run-time location of the workpiece, a trajectory that will avoid the troublesome manipulator postures. Sensor-controlled gross motions can easily lead the manipulator into postures for which multiple solutions and singularities occur.

Since conveyors are not used very much in aerospace manufacturing, the work-station controller will probably not have to be able to track moving points or follow around moving contours. However, software techniques that are adequate for these tasks and that can run in today's minicomputers or microcomputers exist. The main advantage of a work station in which there is no need to track moving parts is that arrival-time requirements can probably be eliminated. This will greatly simplify the portion of the work-station software that controls concurrent processing.

On the other hand, contour-following at a specified or sensor-controlled speed along that contour is necessarily time-critical. Even if the work station must support such activity, a good general principle to follow is to try to arrange for every piece of equipment in the work station to be totally asynchronous in its interactions with other equipment. Each piece of active tooling should be able to wait indefinitely for any other one to perform its function. The work-station control computer alone should detect when any piece of equipment is taking an unusually long (or short) amount of time to do its job. If so, the computer alone should institute corrective or diagnostic action. A work station designed in this way will be much easier to set up for new jobs and will be much more reliable. This comes about because critical timings and race conditions will have been eliminated. Such conditions can be difficult to observe or reproduce so that they can be corrected; consequently, they make debugging very difficult.

5.2.4.3.2 **Template-Following** - Templates are important items of tooling that can reduce the amount of manipulator accuracy and stiffness needed in order to accomplish a precise point-processing or line-following task.

If an end effector touches a workpiece, large forces may be applied to both. Manipulators are notoriously compliant mechanisms in comparison with traditional machine tools, so large contact forces at the tool tip can easily result in sizeable position errors, at least by aerospace standards.

Templates are simple mechanisms for obtaining accuracy. Templates for point-processing applications must constrain the end-effector position in two translatory degrees of freedom along the surface of the workpiece. Templates for line-following applications have to constrain position in only one translatory degree of freedom, namely the direction normal to the contour or path being followed and parallel to the workpiece surface. In either case, the template may also be required to constrain the orientation of the end effector in one or more rotational degrees of freedom.

The work-station controller must allow the template to constrain the end effector's position, yet must still be able to move the template toward the surface (in point-processing tasks) or along the contour (in line-following tasks). This requires that the tool be held with a stiffness that is different in different directions (anisotropic compliance). The work station can achieve this kind of control in several different ways.

The simplest way is to attach the tool to the manipulator with a properly designed compliant mounting. This mounting should be stiff enough to exert as much force in the required direction(s) as is required for the particular process that the end effector performs. Contacting tools, such as drills and routers, for example, should be pressed against the workpiece. The mounting should also have sufficient range of motion to accommodate to the largest variation in the shape of the workpiece that is likely to be encountered. Either springs or pneumatic actuators can supply the required component of compliance. The tool mounting should usually be very stiff in one or more directions; the number of directions will depend upon the specific tool. For example, in routing, the router mounting should not allow the tool axis to rotate about the line tangent to the edge of the template; otherwise, the tool will overcut or undercut the edge.

Another more complex template-following technique is to sense the forces at the tool (perhaps with a sensor mounted back in the manipulator's wrist) and then use that information to control the motion of the tool. One rather general algorithm is to form a six-element vector from the three force and three moment readings, and multiply that vector by a six-by-six compliance matrix. Taking the resulting six-element vector as a velocity at which to move the tool (both in translation and rotation) can produce a variety of useful automatic edge- or surface-following behaviors. Taking that vector as an amount by which to displace and rotate the tool from a nominal position can result in automatic accommodation motions useful for fitting parts together. For successful use of this method, the manipulator must be able to accelerate and decelerate fast enough to stop within the end effector's compliance distance after it contacts a solid object (otherwise, either the end effector or the object must break). In order to prevent oscillations (bouncing along the template, for example), the servo system should be able to sample the sensors and update the manipulator's set points before the end effector deflects through more than a small fraction of its compliance range.

For example, suppose that the end effector can comply only 0.1 of an inch (2.5 mm) in any direction and that it contacts a workpiece while traveling toward it at 2.0 inches (5 cm) per second. At that speed, the end effector will collide with the workpiece and damage it in only 0.05 second. The manipulator will have to decelerate the end effector at a rate of approximately at least 20 inches (51 cm) per second to stop it within the compliance distance. If the force sensor is read less frequently than every 0.05 seconds, then the damage can occur before the system even realizes that contact occurred. Any practical servo system must be able to react much faster than this. A sampling rate, one order of magnitude faster (say, every .005 second), might be adequate.

Making the required servoing calculations at such sampling rate can present quite a challenge to a microcomputer. A typical force-controlled servoing task might require the microcomputer to perform the following computations during each sampling interval:

1. Read strain-gage signals from six or more wrist-mounted strain gages, then apply scaling and offset calibration factors in order to obtain stress values.

2. Filter out noise in the stress signals resulting from vibrations and electrical interference.

3. Compute three components of force and three of torque measured at the wrist.

4. Transform that force and torque into the force and torque that would have been measured at the tool tip.

5. Compute the tool motion (displacement, velocity, acceleration, etc.) required in response to the tool-tip forces. This will depend on the tool and on what the work station is trying to do with it.

6. Compute the motion that each manipulator joint should make in order to produce that tool motion.

7. Send the manipulator appropriate commands to make the joints move in that way.

8. Wait until it is time to read the sensors again.

The potential complexity of servo calculations like these may make the design of special hardware for some computations and distribution of other computations among several small processors worthwhile.

A third approach to template-following is called the buried-set-point method. Low-gain (soft) servos are used in this method to apply force along the edge of the template. In order to cause the force to be applied, the programmed path of the tool is positioned parallel to and buried (perhaps an inch) below the surface of the template. In operation, as the tool moves along the surface, an error which is sensed by the servo control circuit exists between the actual tool position and the programmed position. To reduce the error, the servo circuit causes a force to develop against the template proportional to the instantaneous error in the position of the tool. The gain of the servos must be set low enough to prevent the manipulator from pushing the tool too hard against the workpiece and damaging it. (A high-gain or stiff servo system using a compliant end-effector to absorb the excess force can achieve somewhat the same effect.)

5.2.4.3.3 Aerospace Processes - Pick-and-place tasks, drilling, spot welding, and stud welding require only the ability to go to a point. Automatic calibration, indexing, and end-effector orienting procedures usually require go-to-a-point motions in which the

location of the point is determined by a sensor. In many aerospace applications, the manipulator will only have to bring a point-processing tool to some given position with respect to the workpiece and then just hold it stationary while it performs the function. This is particularly true of the smart tools described above. If a particular manufacturing situation will permit the drilling tool to attach itself rigidly to the workpiece (or to the jig that holds the workpiece), once the manipualtor has gotten it into the correct position, the manipulator-control problems will become even less difficult.

Routing only involves following a contour at any convenient speed that will produce accurate results. Seam welding and spray painting require contour-following at a specific velocity. Most two-handed aerospace manufacturing tasks, such as two-sided riveting, can be performed asynchronously. Two-handed manipulation of metal or fiber-composite sheets may be the exceptions. In such tasks, the motions of the two hands must be coordinated. In order to prevent buckling of the sheet due to transient servo errors that reduce the distance between the grippers, one of the hands could be designed or operated to provide enough compliance to maintain constant tension on the sheet.

Template-following techniques are applicable to tasks that require extreme accuracy or tasks in which large forces arise from contact between the end effector and the workpiece. These tasks include point-processing tasks, such as drilling, countersinking, and riveting, as well as line-following tasks, such as routing, grinding, and deburring. These techniques are probably not applicable to area-coverage tasks because the templates would probably have to be impractically complex structures, and most area-covering tasks do not require great accuracy in following a contour.

Overlapping of coverage from adjacent passes in an area-covering application may or may not matter. In applications such as shot peening, sand blasting, or spraying cleaning fluids, overlapping does not matter. Overlap does matter in spray painting though. If it causes too much paint to be deposited in places, puddles and runs will form and will mar the finish. However, if successive passes of the spray gun do not overlap to some extent, the surface will not be completely covered.

At first glance, using a point-processing tool on a moving workpiece might seem to be as difficult as using a line-following tool on a stationary workpiece, but this is not always the case. Consider the important class of tracking-and-acquisition tasks, such as picking up a workpiece from a moving conveyor. In such a task, the trajectory of the end effector -- usually some sort of gripper -- only has to match the motion of the workpiece long enough for the end effector to operate. This procedure can be a very short time and distance. Picking up objects with a magnet or suction cup rather than with a gripper may be easier if the gripper has fingers that must be placed carefully around the object before they can close. Furthermore, the work-station controller often has the freedom to pick up the workpiece anywhere along the conveyor line, not just at a specific point. This freedom almost always makes the control problem simpler.

In practice, manipulator dynamics will probably pose the most problems in any velocity-contolled contour-following tracking or multihand coordination applications. Some work has been done on compensation for these dynamic limitations and on meeting arrival-time constraints. Software for fast visual tracking of moving objects is in the research stage at present (Reference 42).

5.3 CONTROL STRUCTURE*

This section provides guidelines for development of an industrial robot control structure. Topics discussed in this section include the control issues of industrial robots, the requirements a control system must satisfy, control architecture, and the ICAM hierarchical structure.

5.3.1 Control Issues of Industrial Robots

This section addresses industrial robots that have some form of servoed actuators. Aspects of the control structure design are applicable to limited sequence (non-servoed) robots but will not be elaborated on at this time. Robots employing servoed actuators can potentially perform a large number of tasks now done by human workers. Since servoed actuators can be commanded to go to any position along their path of travel, these robots, in theory, can accomplish almost any function. Speed, accuracy, and rigidity requirements for some applications exceed the inherent capabilities of the servoed mechanical system, but, in general, this type of system can effectively accomplish a large number of tasks if the proper control is provided.

Industrial robots are presently treated as semihard automation, i.e., performing repetitive jobs in long production runs and working with parts that are rigidly constrained and accurately positioned. This is directly related to the difficulty in programming new tasks and the inability to interact with sensory feedback data that would inform the system of misalignment of parts and error situations in the work environment. Further, since teaching a new task to the robot is done by leading the robot through the required steps, the work station facility is unproductive while it is being used for this teaching operation.

All of these impediments to the more effective use of robots result from the limited control systems now in use. Most of the present control systems are no more than tape recorders. Critical points along the task trajectories are stored in sequence by reading in the actual values of the joint position encoders at each location. In order to perform the task, these points are played back to the robot servo system. Some systems allow branching to alternate stored sequences on the basis of some external signal. In this way, the robot may handle a variety of workpieces. However, the time-consuming and tedious teach process involved in recording these points and an inability to modify trajectories on the basis of sensory data will always be inherent problems with this method of control.

Some robot manufacturers have replaced the described wired-logic controllers with general purpose computers. A computer offers the potential of providing the necessary control capabilities that enable industrial robots to become truly flexible automation equipment. Enhanced man-machine interface is possible as is efficient processing of sensor data. Using a computer to process sensor data obtained from the robot and to interpret task-oriented commands provided by the user, the robot system can exhibit goal-directed behavior where the goal is task completion rather than simple trajectory motion. However, concomitant with less tedious and faster programming ability and a sensor-interactive behavior is an increase in the complexity of the underlying control

---

* Subsections 5.3.1, 5.3.2, 5.3.3, and 5.3.4 adapted from "Robotics Support Project for the Air Force ICAM Program," Second Quarterly Interim Report, National Bureau of Standards, April 1979.

structure. Since this control structure is to be implemented as computer programs, the software design and programming techniques become crucial. This will be a large, complex software system, and the method of design and implementation must allow for changes to be made easily and for the system to be maintainable, comprehensible, and reliable.

5.3.2 Control System Requirements

An effective, flexible robot system must be able to handle inputs from two distinct sources; one set of inputs will be the commands that will define the task to be done, and the other will be the sensory data that will describe the environment. A well-defined interface between the user and the system is required to enable quick and easy task specification. In addition, the system must interact with the environment through sensors and use this information to modify its behavior.

5.3.2.1 Robot/User Interface

Large amounts of detailed information describing all the aspects of the task to be performed must be supplied to the robot system. The control system interface specifies the structure and format that enable the user to present all the various types of information required. The user must supply the information easily. To simplify the user's job of communicating with the robot system, the interface should allow a high-level task description language and the ability to symbolically specify data points to aid in separating the concept of task description from the various types of data that are assigned numerical values based on the particular robot, sensors, work station, and workpieces used. Extending the capabilities of the control system should not be a difficult task.

5.3.2.2 Multilevel Interaction

The present communication interface between the user and the machine involves leading the robot through the correct sequence of actions. A more desirable interface would be one that allows the operator to tell the robot what is to be done rather than to teach by doing. For the robot system to be a truly flexible automation tool, this task instruction must be fast and easy to accomplish. Because of the large increase in complexity of the control structure that this requires, at least two different levels of user interaction with the control structure, as shown in Figure 25, are suggested. One level, tied into the inner workings and architecture of the control structure, is at a very high skill level. This level involves programming the detailed steps, procedures, and algorithms required for interpreting the different input commands, interacting with sensory data, generating trajectories and error recovery procedures, accessing data bases, etc. The other provides the fast, efficient, simple programming interface whereby an operator (task programmer) can describe a task or procedure at a symbolic level with much the same ease as he could to a human worker. These high-level commands are then operated on by the control structure set up by the first type of programmer (control system programmer).

To have only a single type of user interface would result in tedious and time-consuming task specification since programming would have to be at the complex, detailed level, involving adding, deleting, and modifying control structure algorithms. In addition, the skill level required of the programmer would be that of a computer scientist.

There are strong parallels to this multilevel interaction in the human work force. Teachers in the forms of parents, school instructors, peers, etc. provide detailed

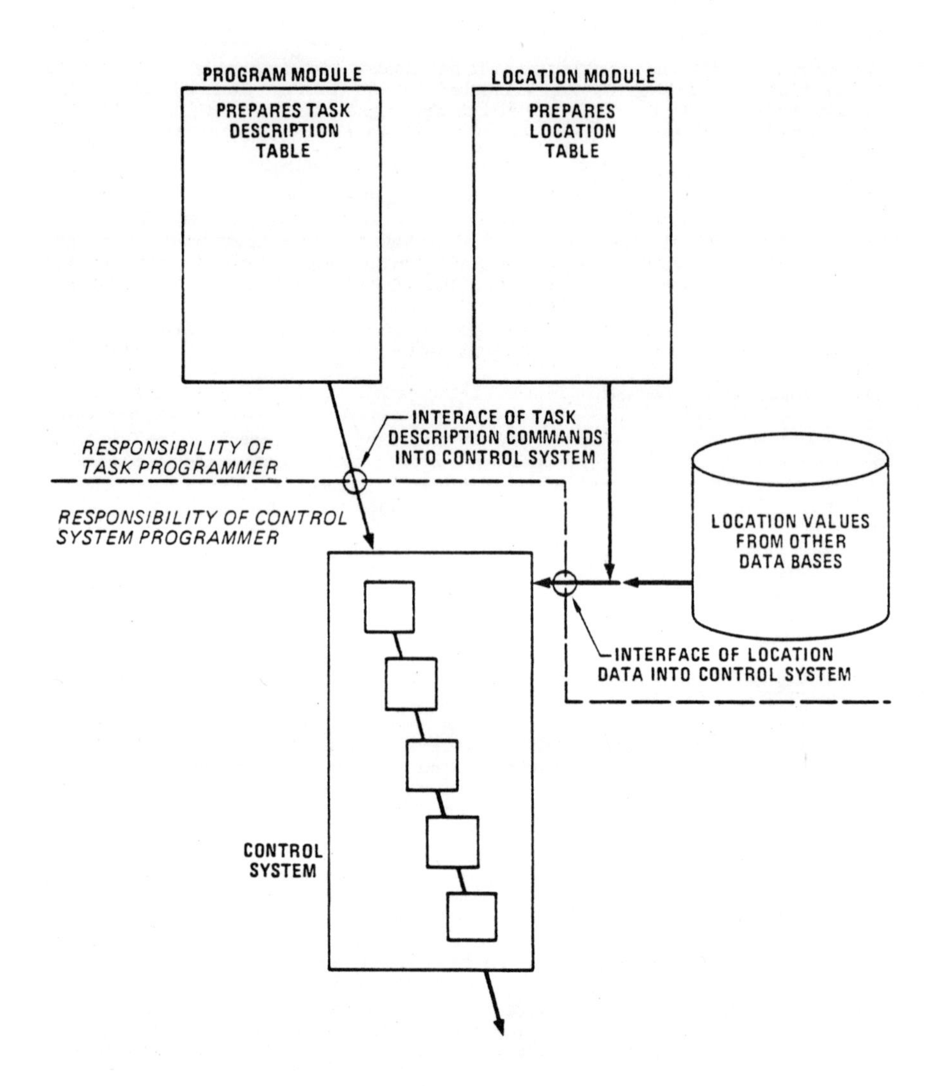

Figure 25
LEVELS OF USER INTERACTION WITH THE CONTROL SYSTEM

instruction. A person learns primitive functions like walking, talking, manipulating, writing, reading, handling logical and arithmetic functions, and learning detailed sequences of operations to which symbolic names are assigned. After many years of this tedious and time-consuming detailed programming, the human worker has knowledge (a data base) of many types of procedures. A supervisor can instruct the worker at a task description level. Only a high-level symbolic description of the desired task or goal has to be given. The worker can be instructed, for example, to screw the top plate of a carburetor onto the base subassembly. This, then, is the input task command. The worker does not have to be instructed as to how the top plate is to be picked up, how to pick up the screwdriver, where to put the screw, which direction to turn the screwdriver, etc. All of these functions have been learned (programmed) previously and are recalled from experience to enable the worker to do the job.

Different levels of interface to the robot control structure could provide the operator (task programmer) with the same kind of high-level task specification. This person would program a simple high-level procedural description of the task. The control structure would decompose this into the correct sequence of detailed steps by execution of appropriate algorithms. This control structure is not generating actions by any intelligent decision-making process; this is a totally deterministic system where all of the responses and courses of action have been programmed in by the control systems programmer. This person has defined the set of possible input states (i.e., the allowable input high-level task description commands, the possible sensory data input values, the error conditions that will be attended, etc.) and the possible set of resultant output states. The robot system is then programmed (educated) by the control system programmer writing algorithms to define which responses will be evoked for which set of inputs. This form of preprogrammed goal-directed behavior is used by a number of groups. If a response to a certain set of input conditions is not programmed, the control system can neither decide on the proper output nor learn the correct response. All of the intelligence to respond to the environment must come from human intervention in the form of previously programmed functions.

Thus, given the example above of screwing the top plate onto the carburetor, the control system would not decide on which orientation to place it or where to grip it or which screws to put in first. This type of planning and decision-making falls properly into the area of artificial intelligence and is still very much a research problem. The control system described here is a very different, totally predetermined system. All of the detailed actions and their sequence will have been programmed into the algorithms by the control system programmer; the part locations, gripping points, etc. will have been specified in the data base. However, within the set of defined input states, this deterministic system will exhibit goal-directed, adaptive behavior, resonding to sensory feedback to modify the robot's motions in real time so that the task will be accomplished in spite of perturbations in the environment.

5.3.2.3 Independence From Data Bases

Another feature supportive of the concept of fast, efficient task description is the accessing of data bases for location point values. If there already exists a data representation of some of the trajectory points for the particular task, then it would be advantageous for the task programmer to have the ability to access these points readily instead of having to duplicate this data in a teach operation. For this feature to be realized, advanced development of the concepts of off-line programming, automatic indexing, and coordinate transformation will be necessary.

The control structure should allow the task programmer to enter a coordinate description of points or use the teach method if desired. However, once these points are in the data base of a robot, they should be maintained in a general enough format to be usable by any other robot. These values should be stored as some relative coordinate reference frame values, not as the joint values of a particular robot. Thus, if one robot is replaced by another at a work station, the same data base of points should be usable and independent of the particular robot.

5.3.2.4 Task and Data Independence

The concept of robot independence of the data base of location points requires the separation of the task description and the data. This can be accomplished by the symbolic naming of locations which will be assigned numerical values from the location data base before or at execution time. The symbolic naming does much to ease the task programmer's job. Providing named variables like vise, drill, or hole not only makes the task description more comprehensible but also relieves the task programmer of the burden of supplying numerical values when he should only be specifying a procedure.

5.3.2.5 Work Station Independence

Separating the descriptions of procedures from the location data base allows a task to be programmed relatively independent of the particular work station that will execute it. The task description, since it is a specification of a procedure, will not change from work station to work station unless there is something different about the station such as parts arriving palletized instead of randomly oriented on a belt conveyor.

Robot independence can be extended to an even deeper level from the control system programmer's point of view by programming as much of the control system, sensory processing, and error-recovery algorithms as possible in a form independent of the robot, the work station, and the computer hardware. The advantage of this approach is the transferability of a large part of the control structure to each work station. This reduces to a minimum the amount of duplication in creating the control structure for each robot and permits the control system programmer to expend a greater effort on improving a generalized control structure instead of regenerating identical control algorithms for each new robot.

5.3.2.6 Extensibility

Due to the desired general nature of industrial robots, all the possible control algorithms, input commands, sensory error-correcting techniques, etc. cannot be fore seen. Therefore, to be effective, the system must permit additions or deletions of functions, as well as changes in existing functions to be made easily. If the control structure is well designed and modular, it should greatly enhance the ease and speed with which the systems programmer can incorporate changes while keeping the high degree of reliability that is an absolute requirement of the system.

5.3.2.7 Adaptability

At present, the goal state of an industrial robot is to go to certain prerecorded points in space. The goal should be the completion of some high-level task, such as drilling holes in a wing skin, positioning a subassembly for proper riveting, etc. This change in the level of the assigned goal from specified points in space to a procedural task requires a closed-loop control system. Measurements of the relevant parameters in the work environment such as the position of the drill guide bushings, the positions of a support strut on a subassembly, the absence of rivet in the drivematic, etc. must be made. This sensory data must be processed in a form suitable for determining branch conditions to algorithms providing real-time trajectory corrections, alternative procedures, or error reporting. In this manner, the control structure becomes responsive to perturbations in the work environment and is able to accomplish high-level tasks while adapting to varying situations. This sensory-interactive capability relaxes the requirement of precise positioning of parts to the robot. Parts can come into a work station with somewhat random orientation if a sensor system can detect the orientation of the part and then correct the robot's pick-up trajectory to accommodate. This ability also offers the possibility of increased reliability since a set of responses to possible error conditions is programmed into the system. As errors occur, the system reacts and continues to be productive. If a new type of error condition occurs for which there is not programmed response, the control system programmer codes in a corrective algorithm.

The effectiveness of this incorporation of new responses is, of course, dependent on the availability of the programming characteristics described in the previous section. Adaptability, therefore, implies accommodating to the environment, which in turn implies processing some feedback data concerning the state of the environment. This will encompass the interaction of the control of the robot with inputs from all types of external objects, sensors, active tooling, material transport systems, machine tools, instrumented jigs, other robots, etc. This interaction must be constrained to provide effective responses within a minimal time frame.

5.3.2.8 Reliability

Implicit in the discussion of an effective system has been the notion of reliability. Especially with capital intensive equipment such as industrial robots, reliability is essential to their productivity since their payback is dependent on full utilization. A control structure that provides a well-defined user system interface and is responsive to the enviroment but is not reliable is useless in the manufacturing world. The overall design or architecture is important to develop a reliable control system.

A control system architecture should provide the control system programmer with the framework necessary to implement the above features in the simplest manner possible and in such a way to allow the system to be easily extendable. The system architecture should also provide the underlying organization to allow the control system programmer to view the overall structure and interactions of the entire control system; to keep the system understandable and comprehensible; and to retain a grasp of the big picture. This helps to prevent the unnecessary introduction of complexities and unknown states into the system. For example, if the visual processing of camera data were intricately interwoven with control algorithms and a new sensor was to be incorporated, a large number of patches to the control structure would be required. This, of course, is a guaranteed method of producing an unreliable system. Thus the goal of reliability is fundamentally impacted by the architecture of the control system.

5.3.3 Architecture for a Control System

A hierarchical structure has been chosen by a number of groups (1, 6, 9, 11, 12) as the fundamental framework on which to build a control system. The hierarchical structure forces a decomposition of high-level tasks by the top level into sets of procedurally simpler subtasks, which become goals to the next lower level. Each of these is decomposed into sets of yet simpler subgoals so that the bottom level provides the detailed steps required to accomplish the task. Therefore, each level of the hierarchy receives input goals of equal procedural complexity. The function of the level is to interpret these inputs and produce outputs that will be inputs to the next lower level.

5.3.3.1 Task Decomposition

The top level in the robot control hierarchy will cause decomposition of a high-level task command, such as LOAD PART INTO VISE, into a set of subgoal commands that are equivalent in procedural complexity to each other. In this example, each of the resultant subgoals (GO TO PART PICK UP, GO TO VISE INSERT, GO TO SAFE POSITION) are of the same degree of complexity because each represents a single trajectory motion followed by an operation or a halt. The next lower level would accept each of these subgoals as its input commands and generate sets of output commands. As an example, GO TO PART PICK UP, might generate the set of outputs: ACCELERATE, MOVE AT CONSTANT VELOCITY, DECELERATE, APPROACH, GRASP. These would be the input commands to the next lower level, and each is the same degree of complexity in that they are either trajectory segments or operation primitives. Figure 26 shows the decomposition of these commands.

Thus, this progressive decomposition of input goals by each level into sets of subgoals of equivalent procedural complexity has structured the functional requirements of the control algorithms. The set of inputs and outputs for each level is specified by the input goals and output sets of subgoals. The system programmer then codes into the algorithms the desired functional relationship between the input goal command and the set of output subgoals to the next level. The function of the control algorithms for each level is to provide the desired output subgoal condition given a specified input state or goal.

Each level within the hierarchical control system, threrefore, represents a functional relationship between an input state and its resultant output state. There are other inputs that can be considered in addition to the input command goal. Status reports from the level immediately below, reporting on that level's effectiveness in completing its assigned goal (one of the output subgoals from the present level), can also be used as input. Further inputs may arrive at each level in the form of feedback from sensory processing algorithms.

A change in the state of any of these input conditions will result in a change in the output state. The output state can consist not only of output command subgoals to the next lower level and a status report to the next higher level indicating successful completion of input commands or the occurrence of error conditions, but also of requests and predictions to the sensory processing algorithms.

5.3.3.2 Sensory Data Input Processing

Processed sensory data becomes an input to the highest level of the control structure that will generate a change in the output state as a result of a change in this data. Figure 27 shows the interactions of inputs and outputs for a level in the control hierarchy.

For example, if a vision system were to be incorporated where the degree of sensory processing would result in data indicating the presence or absence of a part in its assigned neighborhood, the hierarchical control structure would greatly facilitate the identification of which level this data should be a part. It should not be an input to the level that generates different primitives and trajectory segments. This complexity of processed data should have an effect greater than just choosing a different trajectory segment. A totally different trajectory to an alternate location for another part may be required. In this case, the sensory data should input to the level that outputs entire trajectories.

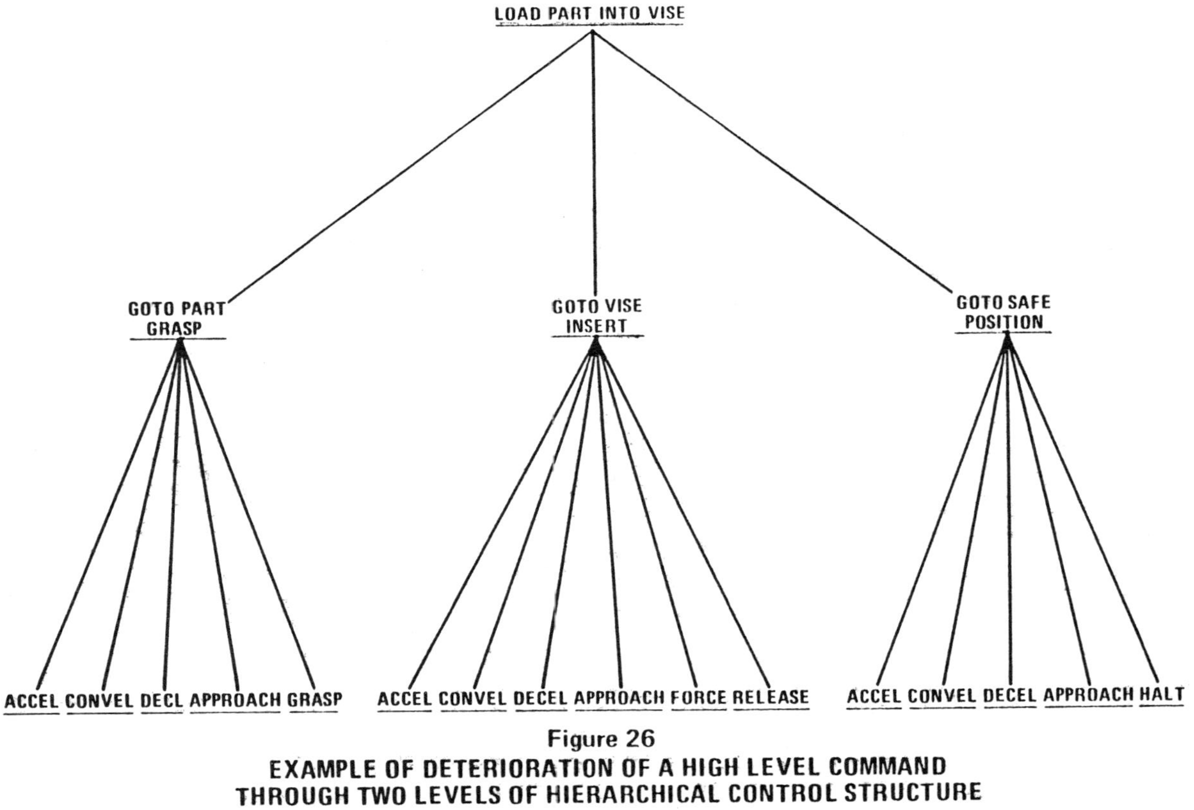

Figure 26
EXAMPLE OF DETERIORATION OF A HIGH LEVEL COMMAND THROUGH TWO LEVELS OF HIERARCHICAL CONTROL STRUCTURE

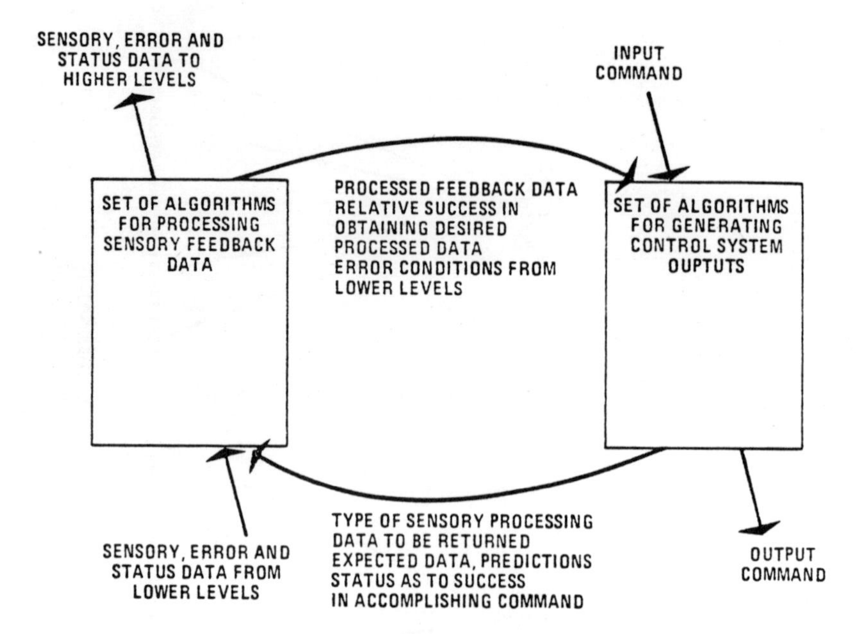

Figure 27
DETAILED VIEW OF THE TYPES OF INTERACTIONS BETWEEN THE SENSORY AND CONTROL HEIRARCHIES AT ONE LEVEL

Thus, the structure of the hierarchical control system, together with the response of the system to the processed sensory data, specifies at which point it should be incorporated into the system. The inverse of the above is also true. The degree of procedural complexity of the outputs of a level help to specify the amount of sensory processing required of the input. The sensory input into a level is only processed enough so that it provides sufficient information for the control algorithm at that level to branch to another output state.

5.3.4 Advantages of Hierarchical Control

The hierarchical architecture of the control system, by providing the framework for the procedural decomposition of an input task and the specification of the above described inputs and outputs at each level, has provided a number of advantages.

The decomposition technique of a hierarchy aids in generating the desired characteristics described in Subsection 5.3.2. The input to each successive higher level in a hierarchy is a more procedurally oriented task description. The input to the top level of the hierarchy quite naturally defines the task programmer's interface into the control structure. The writing of the algorithms to decompose and execute these procedural task descriptions and to interact with sensor data, i.e., the hierarchical control structure itself, is the responsibility of the control system programmer. Thus, the hierarchical structure has helped define the two levels of user interface.

The decomposition of the task into sets of simpler, procedurally equivalent subgoals in a hierarchical framework aids in defining the fundamental requirements of control algorithms at each level. The sets of inputs and outputs for the control algorithms are specified, and the system programmer merely has to code in the desired functional relationships between them at each level. The overall system has been kept comprehensible since it is structured on a framework that identifies the location in the whole control structure for each level of control algorithm and thus by its very structure displays the relationship between the algorithms throughout the control system. This greatly aids in identifying and specifying the data input interfaces and the interfaces between algorithms and control levels. For example, the highest level in the hierarchy where the procedural decomposition process requires the actual numerical specification of the location point values for its output commands to the next lower level then determines the place in the control system where this data base must interface. The structure of the data is defined by the format required for interfacing the data to this level.

The hierarchical structure of the control system has, therefore, impacted greatly the identification and specification of both the task programmer's and control system programmer's interface and responsibilities. This structure has simplified the task of separating the control algorithms into functionally distinct levels within a comprehensible framework and has aided in the identification of interfaces between the control system and data bases and the specification of the structure of the data. The structure has enabled identifying where sensor data can most effectively be incorporated and what level of sensor data processing is required at what point in the control structure. All of these benefits combine to make the system extensible.

5.3.5 ICAM Defined Structure

An objective of the USAF Integrated Computer Aided Manufacturing (ICAM) program is to produce systematically related modules for efficient manufacturing management and operations in the aerospace industry. The enabling program philosophy

is in harmony with hierarchical structuring which recognizes operating stages of increasing responsibility, complexity, and susceptibility to computer enhancement. As indicated in Figure 28, the stages are categorized as process, station, cell, center, and factory, each having its own software and hardware needs and operating modes.

A manufacturing process is a single operation or set of operations carried out by a person or machine not aided by an external hierarchy of program-driven circuitry or computerized software. Processes are primarily controlled by a person or a station controller. The highest level of control for a process is a cell.

The station is the lowest level of automated control and is composed of sets of manufacturing processes under the control of software resident in or under the direction of the respective station. Stations control processes and operate in real time. Stations are controlled by cell controller software.

A cell is the automated control of one or more stations to include material handling and may include a single process external to any station control in the

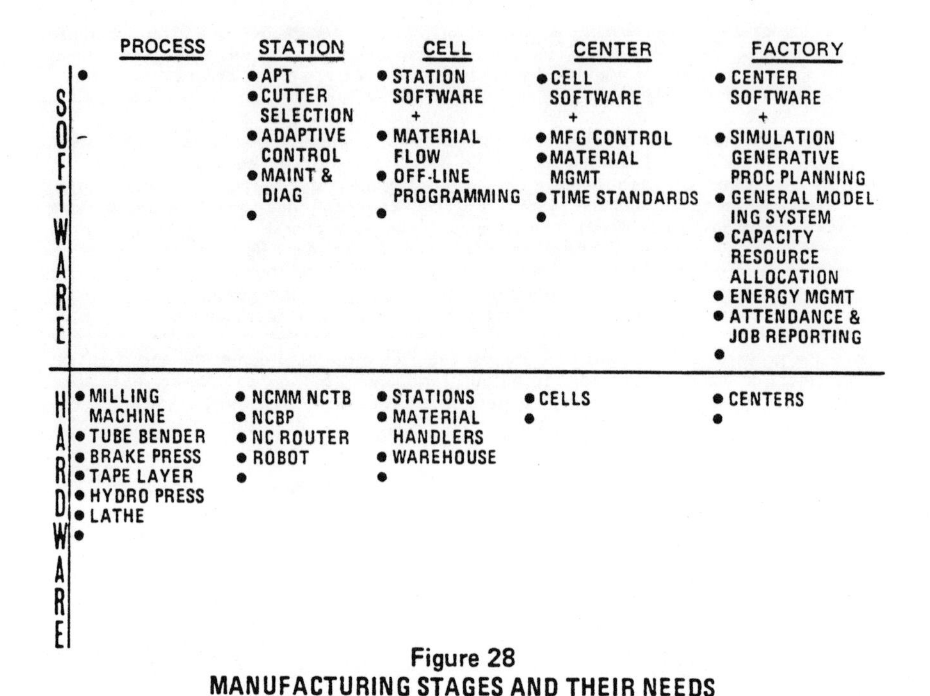

Figure 28
MANUFACTURING STAGES AND THEIR NEEDS

respective cell. A single station under cell control would have to be accompanied by a process not under station control. Cells are controlled by center software. At this level interactive design activities are supported, large data bases are managed, and service is provided to the production engineers for development of task programs to be carried out by the work stations.

A center is the automated control of two or more cells. A center may include a single station external to any inclusive cell. A single cell under center control must be accompanied by a single external station. Centers are controlled by factory controller software. The work center is primarily responsible for job-shop scheduling and related activites. The center is tasked to maintain a large data base that describes the present status of the work stations under its control.

The factory is the automated control of two or more centers; however, it may include a single cell external to any inclusive center. A single center under factory control would require accompaniment of a single external cell. Factories are controlled by management personnel and policies.

The work cell and work center communicate with each other about their present status and the capabilities of the work center as a whole. The work cell transmits to the work center detailed instructions for production activity in the work stations. The work center mainly performs a store-and-forward operation, passing the detailed instructions to the work station as they become ready for new jobs. The center is also responsible for job-shop scheduling over the set of work stations in its domain. The work center can also provide various simple services to the work-station computers, such as reloading them after unrecoverable failures. The work center and work station primarily communicate information about one specific manufacturing task, e.g., instructions for drilling rivet holes in an airframe panel. In this situation the work center would pass the detailed instructions obtained from the work cell to the work station.

First results of the ICAM program have been integrated into the lower hierarchies of the architecture of manufacturing, that is, at the process and station levels (building in a "bottom-up" fashion). The lower levels have been advanced in performance primarily by hardware developments under ICAM, while the integration required at successively higher levels will depend increasingly on software.

At General Dynamics, the concept of an Automatic Trim Cell was developed. The cell was defined as an area within the Sheet Metal Work Center that consists of a grouping of subcell or work stations that could be integrated by a material-handling system. Each work station would consist of a robot, multiaxis part presentation system, and end effectors - all with integrated computer controls, and ancillary equipment that will enable the cell to process a majority of the sheet metal parts currently processed in the hand rout area. Figure 29 illustrates the concept of an Automated Trim Cell consisting of four robotic work stations. Two work stations are limited to small sheet metal parts. The other two stations contain larger computer-controlled multiaxis part presentation systems that will enable the station to process large, highly contoured parts that cannot be processed in existing equipment.

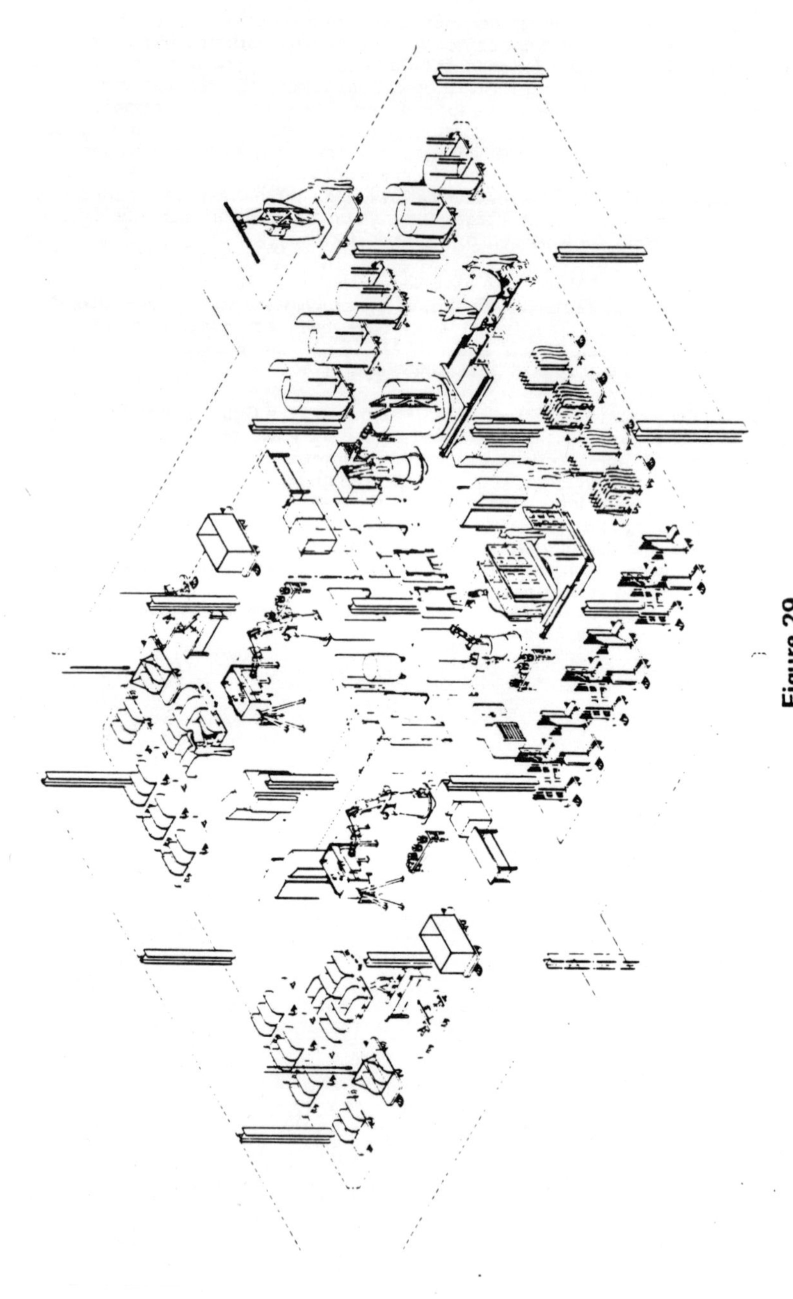

Figure 29
ROBOTIC TRIM CELL

-6-

Application Information

For years robots have been applied to manufacturing tasks. Their uses span many industries from aerospace to the foundry. The process of implementing robotic technology can be more efficient if the results of those applications and research already accomplished is reviewed.

The capabilities of robots are limited, and for a successful application, the proper selection of the robot itself is only one of the important application ingredients. A total system approach should be used. This section will address some of the additional information that will be useful in designing and implementing a robotic system.

6.1 CURRENT AREAS OF APPLICATION

The spectrum of robot usage is very broad. Because of the advances of the state of the art in robotics and in computer technology, the potential applications are almost without limit. Six categories of robot applications are identified here.

- **Pick and place** - This is the utilization of the robot in moving objects from one place to another and positioning materials in the manufacturing process. Tasks include material handling, grasping, transporting, and heavy-duty handling.

- **Machine loading** - In this application, the robot is combined with another machine and accomplishes the material loading and tool changing. Examples are robot loading of numerically controlled milling machines, lathes, and automatic presses.

- **Continuous path** - This application involves a process in which a precise rate of motion may be required. Spray painting and welding are common examples. In both, the motion of the robot must be synchronized with the rate of application or speed of the associated process. Some attempts have been made in using robots to apply epoxy resin for composite layup. In this example, the robots are used to spray the resin between the successive layups of graphite broadgoods.

- **Manufacturing Processes** - A robot for this application is one which is dedicated to cutting, forming, finishing, or otherwise processing materials for manufacture. In the aerospace industry, robots are being used to drill and rout aluminum sheet metal and graphite composite panels. This application generally requires extensive tooling design work as described in previous sections of the guide.

- **Assembly** - This is largely still a research area and most of the current literature in this area is from research programs. A robot for assembly would be designed to mate or fasten parts together into an assembly. Assembly applications characteristically require a relatively more articulate robot with high-level sensory feedback and control capability and complex tooling and parts feeders. Vision acquisition and force feedback systems that will provide better adaptability are areas receiving much attention in assembly applications.

- **Inspection** - These systems appear very similar to assembly systems in that they may require precise control. A robot for this application will generally either position material, parts, or the precision measuring instrument itself for the purpose of checking some aspect of the parts or material. Examples of components used with robots for inspection are television cameras, linear diode arrays, fiber optics, lasers, and photoelectric control modules.

Information for a particular application can be obtained from several sources. Many articles describing specific applications have been published. Many of these are listed in Appendix B, List of Current Literature. Additionally, many robot manufacturers have extensive information on applications for which their products have been or can be applied to.

6.2 IMPLEMENTATION FACTORS

The question of whether or not to implement robotics technology usually arises from a realization of a problem in the flow of production where robotics technology offers a possible solution. Other solutions to the situation may be available, and a justification analysis should be performed to determine which approach is most desirable with all factors considered. If the analysis indicates a robotics solution, everyone who is to play a major role in the implementation process must be familiarized with the technical approach chosen.

Upper management needs to know what the system can do for the company. These people are the ones who will decide the basic policy toward robotic technology and who will take most of the risks. Therefore, all data, the advantages and the disadvantages, must be presented accurately.

Middle management needs the same information as upper management, but they need more technical detail. Middle management will be responsible for setting up the implementation mechanism once the go-ahead is given, and they must realize the need for training the engineering staff in their new technology. Clear, deliberate planning is essential to successful robotics implementation.

Others to be included in preliminary planning are the plant and assistant plant manager, and operation and engineering managers. They must be fully informed as to how the implementation will affect them. They must agree to take an active part in the implementation, or serious problems or more probably failure will occur. Persons in this management group must display an active interest. Signing an appropriations request prepared by lower management is not necessarily an active interest. At this level of presentation, the abilities and limitations of robotics must be explicit. Special emphasis should be given to the importance of related equipment because whether the robot or its support malfunctions, the robot is usually indicted. Two areas that are often neglected in order to cut costs are training of support personnel and the ancillary equipment supporting the robot. Neglect in these areas could easily mean failure. Watch out for overzealousness and "get that thing into production" haste. Full and complete planning is absolutely essential; this point cannot be overemphasized (Reference 82).

Production supervision should be included in all planning and engineering. Few people have a more intuitive feel for the actual process in question than those who watch and participate in it every day. Use what they know, for this knowledge may save considerable time.

The engineering staff should be fully trained at the manufacturer's facilities. Hands-on experience for this group, and the others too, is highly desirable. The engineers must know the robot thoroughly in order to design an efficient system around it. Much time and money is wasted in false starts and changeovers when details concerning capabilities and limitations are overlooked because of a hasty uninformed approach. The training is well worth the time and effort and should not be neglected.

A successful education phase will create an environment favorable to the smooth implementation of this new technology - a group of knowledgeable engineers and technicians backed by enlightened management. The alternative is an exercise in futility.

The next step begins the work. A thorough analysis of the area of application should be performed in order to determine the functional requirements and technical specifications that will determine the form of the robotic system. Some items that should be considered before choosing a robot for the application are tolerances, work volume, layout, data storage, tooling, environment, and laboratory testing.

6.2.1 Tolerances

For the intended application, a careful and thorough study should be made in order to determine whether the positioning ability of the robot is within the required tolerance. Repeatability is a critical parameter for programs that, once taught, will be run repeatedly for an extended period of time. The maximum allowable error must be determined. The long-term repeatability error of the robot must be less than this value for successful results. If the tolerances cannot be held with currently available robots, the difference may be compensated for by compliant tooling or active sensory feedback control schemes. These alternatives may be developed by the manufacturer or by the user. In either case, reducing positioning errors of a robot below its off-the-shelf capabilities costs money. For quick economical implementation, applications that do not require the robot system to operate at the limits of its optimum capabilities are best.

When positional accuracy is a critical factor, a well-defined and precise reference index is essential. This is especially true when the limits of the manipulator's working range are approached or when off-line programming is anticipated. Robots are generally aligned to a reference plane, and most of them require fastening to a secure base that can be used as the reference. Using a plane or axis on the robot manipulator itself as a reference may be advantageous for a more accurate reference index, not only for the robot but also for the equipment associated with it. This method will eliminate possible alignment errors in the robot mounting and will assist in better defining the workspace of the robot.

An accurate automatic indexing procedure, as discussed in Subsection 5.2.1.5, may reduce the requirement for a precise alignment of the components in the work station. The accuracy with which the position of all components are known will however be essential in the planning of those tasks that approach the limits of the working range of the robot.

6.2.2 Work Volume Selection

The size and shape of the work volume for a particular application are selected through an analysis process in which the application and certain constraints are considered.

The application, whether pick-and-place, manufacturing process, assembly, or inspection, will establish basic criteria and a minimum working range. For example, the selection of a work volume that will accommodate working in a horizontal plane or orienting the wrist in a unique position would be application criteria. Assembling small components is an application that would possibly require only a small work volume; it may also require a robot with a high degree of articulation.

Constraints on work volume selection may be found in two areas: installation environment and in-house design ability. Economic factors also exist and will be discussed in later sections.

The first constraint, the area available for robot installation, may restrict work volume selection because of the nature of the facility or because of management-directed limitations. The available area must be able to accommodate the work volume, associated equipment, parts flow, and maintenance and operator activities. As the work volume of a manipulator increases, the number of things it can collide with also increases. For example, the floor and ceiling of a normal room are within the reach of a Unimate 2000 when it holds an 18-inch-long (45 cm) tool. Arrangement of the work area so that the required work volume is minimized is advantageous provided crowding does not interfere with production. The second constraint, management limitations, could indicate a less-than-favorable attitude toward robotics which should be dealt with early for project success.

The extent or desired depth of tool design for a particular application can be an important factor in selecting work volume. Under certain conditions, fixtures, part positioners, or end effectors may have to be designed quite differently because of the work volume. As previously mentioned, an assembly robot may only require a small work volume; however, this will surely necessitate extensive tooling designed to supply and transfer parts to-and-from and within the work area. Use of a robot with a larger work volume could possibly reduce the tool design task.

As shown, simple quantification does not supply adequate information for work volume selection. The application and design constraints discussed here are correlated with the production facility layout in making the final decision on a robotic system design.

6.2.3 Production Facility Layout

The selection of an appropriate layout requires consideration of the information discussed concerning applications, manipulators, control systems, tooling, and control architecture. There are generally two opposed schools of thought related to facility layout -- the "in-line" school and the "centralized" school. A third approach, the "intermediate," combines features of these two.

- In-Line - Proponents of the in-line school maintain that it will be most cost-effective to arrange several relatively simple robots along a more-or-less conventional transfer line and make each robot do a few simple operations on a part as it comes by. This approach effectively replaces individuals on an assembly line with robots, one-for-one. An advantage of this arrangement is that it can be relatively easy to pull out a malfunctioning robot and replace it temporarily with a person.

- Centralized - The centralized school of thought recommends a few high-performance robots which perform many complex or precise operations on the same workpiece. One advantage of this arrangement is that some duplication of equipment can be avoided; a disadvantage is the inevitable crowding and inaccessibility that result from the number of part feeders and transfer lines surrounding the robot.

- Intermediate - An intermediate approach is to use the in-line approach to put together kits of parts, jigs, and perhaps some specialized tools on general-purpose pallets. A single transfer line would then carry a stream of these kits in and out of a centralized station where a high-performance robot would quickly put the various parts together and create a subassembly. This would allow the centralized arm(s) to operate without the obstruction of part feeders and tool holders, and any cameras used would have a clearer view of the work area.

6.2.4 Data Storage

The amount of data storage required for the application should be considered when choosing a robotic system. The size and number of programs to be run determine the storage requirements. In some installations, the capacity of the internal system storage is insufficient for complete operation. If it is not feasible to remedy this with the addition of more storage capacity, then the next option is generally the increased use of data transfers. With this technique, the operation is divided into parts and transferred sequentially after each part is accomplished.

6.2.5 Tooling

The tooling requirements are at least partially determined by the intent of the application and the performance capabilities of the robot, i.e., tolerances, load capability, etc. Some tools can be purchased from the robot manufacturers while other concepts will have to be developed by the user. Since the tooling can drastically affect the costs, choosing a robot that will allow for the minimization of tooling costs would be advantageous.

6.2.6 Environment

The robot system must be able to withstand the extremes of the environment in which it will operate. Temperature, vapors, dust, vibration, and electromagnetics all must be taken into account and compared to the limitations of the robotic system. This requirement also applies to any peripheral system the user intends to install with the robot system. Generally, the reliability of the entire system will depend on the reliability of each individual critical component. Failure will occur if this aspect is overlooked.

In general, the requirements of the application should be analyzed very thoroughly and compared to available features offered by the various manufacturers. If the available systems cannot meet these requirements, a system to meet some of the requirements should be chosen, and the remaining requirements should be compensated for by manufacturer/user development. Care to ensure that those specifications left unsatisfied can be compensated for in an economic manner should be taken. The goals are minimum total cost and optimum system performance under the existing conditions. A good analysis at this stage will determine the future of the project more than any other single factor.

6.2.7 Laboratory Testing

When the robot arrives, establishing a development laboratory situation is convenient. The robot should not affect the production operations until it has been completely developed and shown to be reliable. This is best accomplished in a laboratory situation. A plan should be prepared for the installation and checkout of the robot, performance studies, development of compensation schemes, fabrication of peripheral compensation systems, tooling studies and fabrication, system integration, testing and debugging, trials, reports, and demonstrations. A realistic plan will help you stay on schedule. Allow time to do the work as well as to report and demonstrate. This stage is the opportunity to ask for time; plan ahead. Asking for and receiving a loose schedule at the beginning and finishing early is far better than overcommitting the group and having to slip the schedule repeatedly. If the robot is not production-ready as initially projected, few people will be sympathetic. A pressure situation will develop and will result in hasty and sometimes disastrous decisions that become irreversible.

A production-ready system formed in the laboratory and thoroughly tested is ready for integration into the factory operations. From this point forward, no fundamental changes in the system should be attempted. Under close supervision, the robotic work should be dismantled and carefully relocated in the factory production area. This relocation is another critical milestone in the implementation process; the robotic system must not be damaged or changed during the move. After installation on the factory floor, the system must again be checked out and debugged thoroughly in order to confirm that the system functions as it did before the move. The operating personnel should be checked out on the system and trained further if necessary. If all checks well, the system is ready for production.

6.3 SAFETY CONSIDERATIONS*

Industrial robots have a remarkably good safety record. No fatalities have occurred in the several million hours robots have operated in factories around the world. However, additional precautions could increase the safety of robots still further.

Industrial robots are helping to eliminate the hazards involved in working in many factory environments. Problems of machine guarding, heat, noise, fumes, and lifting of heavy loads related to metal presses and painting are lessened by these versatile devices. The importance of robots for risk control has been especially great since the Occupational Safety and Health Act went into effect in 1971 (Reference 83). This section includes suggestions of ways to reduce the dangers to workers and equipment and some aspects of OSHA that pertain to robot installation.

6.3.1 Protection Against Software Failures

Though expensive, redundancy offers the best protection against software failure. A double-redundant system can shut itself down when its two components disagree, and a triple-redundant system can use majority logic to override one failed component and continue operation. Both hardware and software redundancy are useful.

Hierarchical and multiprocessor systems can be made more reliable by data redundancy. Messages exchanged between computers should include one or more of the

*This section is adapted from William T. Park, Robot Safety Suggestion, Technical Note No. 159, SRI, International, 29 April 1978, except as noted.

following kinds of redundant information: parity bits, checksums, cyclic redundancy check characters, message sequence numbers, sender and receiver addresses, and error-detecting or even error-correcting codes.

Timeouts are another simple and effective failure test. For example, one could use a timeout in the interface hardware between a robot and its controlling computer. If the computer fails to send the robot interface a keep-alive signal every 100 milliseconds, the interface halts the robot. A special manual or automatic procedure should then be required to get it started again.

Timeouts can also be implemented in software. In a multiprocessor robot control system, one might require regular transmissions between all the computers. The failure of one computer to receive a transmission on time would then indicate a failure in one of the other computers. Specifically, the interrupting computer may have become hung up in a loop, a halt, a noninterruptible state, or it may have suffered a hardware failure. Such transmissions can simply be interrupts since they need not convey any other information.

A status check is a third way to detect software failure. In a status check, one computer sends specific data to a second computer which can tell if the data is self-consistent. The safest way of checking status is to run two identical computers in parallel and compare their actions (possibly with a third, very simple computer). This kind of double-redundant system is very expensive, but one can also make simpler status checks on software operation which are less reliable but still useful. A description of what a given piece of software is supposed to do could form the basis of one type of simple status check. For example, the software in a computer controlling a robot arm should at least try to keep each joint position within the physical limits determined by the design of the arm. Therefore, one simple status check would be to make that computer report the current arm joint positions to a second computer, and have the second computer determine whether the positions are reasonable. This would require only a little of the second computer's time.

6.3.2 Protection Against Hardware Failures

The servo valve is a weak point in a hydraulic system since dirt in the hydraulic fluid can cause the spool to stick in an open position and result in uncontrolled motion of the arm. A precise servo valve is a very complex and expensive device, but it could still be improved in one way. Its cylindrical spool valve could be rotated continuously or back-and-forth around its axis independently of its normal control motion along that axis. This would improve valve operation in two ways. First, the rotational motion would reduce static friction in the valve to zero and make the valve more sensitive to small control signals. Second, it would make it possible to detect a valve clogged by dirt in the fluid since the rotation would stop.

Additional protection against servo-valve failure could be provided by a redundant on-off control valve in the feed line of each servo valve. This would prevent movement of the arm if the servo valve should fail to close.

Sensors that would detect loss of line voltage, pneumatic pressure, or hydraulic pressure, as well as excessive temperature, speed, acceleration, force, and servo errors could be included in the system. Either hardware or software could monitor the signals from such sensors.

Hand tooling and fixtures on breakaway fail-safe mountings should be secured (by a steel cable for instance) to prevent them from traveling more than a few inches.

Redundancy in the individual components of robotic devices and safety systems can make the entire device or system more reliable. Of course, this increases the cost of the system, so it may not always be economically justifiable. Component redundancy can be applied at many levels in a robot system. For example, a robot might have multiple actuators on each joint so that one could fail without making the robot drop what it is carrying. A safety system might have multiple sensors to detect a given hazard condition so that it could continue to operate even if one or more sensors failed.

In order to avoid electrical shock, all robots and their components should be electrically grounded, particularly when welding equipment is part of the robot tooling. Spot welding guns should not be insulated from the robot arm to ensure a ground in case of short circuit (Reference 83).

6.3.3 Fail-Safe Design

Hazard detection sensors, electrical circuits, and other components in a safety device can fail. Equipment that simulates whatever condition the sensor is supposed to detect can be added to guard against this. This equipment would challenge the detection system automatically and periodically and would test for the detection of each challenge. If the sensor should fail to respond to a challenge or if it should respond when no challenge was supposed to be present, then a warning signal would be generated by the safety device.

Thus a fail-safe hazard detector consists of three subsystems: (1) a sensor subsystem for the hazard condition to be detected, (2) a challenge subsystem to exercise that sensor, and (3) a monitor subsystem to watch for any interruption of the challenge-and-response sequence. The challenge and monitor subsystems do not need to be complicated or expensive.

For example, an appropriate challenge to a light barrier used as an intrusion detector would be a small motor-driven vane which repeatedly passes through the light curtain. If the sensor fails to respond when the vane is supposed to be in the path of the light beam, then either the sensor in the barrier has failed or the motion of the vane has been interfered with. If the sensor shows that an object is present in the sensing area when the vane is not supposed to be, then either a real intrusion has occurred, the vane is stuck, or the sensor has failed. Only if the signal from the sensor changes from "safe" to "unsafe" in step with the motion of the vane can we be certain that no obstruction is present and that the safety device itself is operating properly. The monitor subsystem for this example could be quite inexpensive -- for example, a simple capacitor-coupled rectifier circuit which changes the "safe-unsafe" signal voltage into a DC voltage holding open a relay connected to the alarm system. In order to guard against failure modes in which 60-Hz signals from the power line enter the monitor, it should have a filter which would only pass a signal at the same frequency the vane enters and leaves the light curtain (which would be chosen to be harmonically unrelated to 60 Hz).

Three design criteria are important in such safety devices. First, the challenge should not obscure a real danger condition. In the example above, one would arrange for the vane to pass through the light beam many times per second because a real object intruding into the protected space might go undetected for as long as one entire challenge interval.

Second, the portion of the equipment which monitors the response of the sensor to the challenge should be simple enough so that it can economically be made multiple-redundant for maximum reliability. This is necessary because the reliability of the safety devices may thus be constructed from unreliable sensors.

Third, the sensor and challenge subsystems should be as independent as possible of the monitor subsystem in order to ensure the latter's reliability.

- Many manipulators are fail-safe only in the sense that if they lose hydraulic pump pressure, the servo valves can close and trap a volume of oil in all the actuator cylinders; this action supports the arm. The trapped oil leaks past the valves slowly enough that people then have plenty of time to brace up the arm externally if necessary. If the hydraulic lines should rupture, however, it may be impossible to trap oil in the actuators, and the arm may fall onto whatever is below. A desirable safety feature would be quick-acting valves mounted directly on both ports of each weight-bearing actuator. These valves would close in the event of pressure loss and would prevent hydraulic fluid from leaking out of a ruptured hose.

Many refinements on this basic organization are possible, such as having the challenges presented randomly on command from the monitor. Ingenuity and an understanding of the actual hazards to be detected will suggest others.

6.3.4 Intrusion Monitoring

Normally, people should stay out of reach of the arm and any tools it might be holding; some European countries go so far as to require that robots be caged up in case they throw something. Protection could also be afforded by devices, such as pressure-sensitive mats and light curtains that would shut the robot off when anyone came within reach. Without restraints, people will become curious and will walk within the working range of the robot whether it is powered or not (Reference 84).

When people must work close to the arm, they should be required to operate a manual override control which does three things:

1. Overrides the intrusion detection system, permitting the automatic equipment to operate in their presence.

2. Physically constrains the manipulator to move slowly enough so that people can get out of its way if it moves unexpectedly.

3. Enables (but does not turn on) an audible warning device, such as a bell and perhaps also a flashing light.

The computer should be able to detect the operation of the override control, but it should not be able to reset it.

The warning device should operate whenever the arm moves, preferably for a brief interval before the arm begins to move. The warning signal should not be continuous because people quickly become insensitive to constant stimuli.

6.3.5 Deadman Switches and Panic Buttons

Both a deadman switch and a panic button should be installed as protection against the moving arms of the robot. A deadman switch, which must be held to permit arm movement is safer than a panic button because the deadman switch cannot be left carelessly out of reach. However, a panic button should also be provided so that anyone can stop the arm quickly if necessary. An emergency rope may be strung around the robot work area so that a pull anywhere on the rope will operate the panic button.

Training of people who work near the robot should include practice with the deadman switch to develop the correct panic reaction. People can learn to react remarkably quickly in response to an unexpected arm motion.

6.3.6 Workplace Design Considerations

If possible, the system activated by the panic button, deadman switch, or hazard detectors to stop the robot arm should also stop other moving equipment in the area (conveyors, machine tools, cranes). Otherwise, this equipment might run into the arm, and either the arm or the equipment might be damaged.

In designing a system to stop the arm in emergencies, it should be kept in mind that a large arm carrying a heavy load at high speed cannot be stopped quickly without having a severe impact on the arm itself. The longer the arm takes to stop in response to the deadman switch, the less wear on the arm actuators (but conversely, the more likely it is that the arm will hit something before it stops).

A method for protecting the robot operator from hazards has been developed by some aerospace companies and is called a docking-facility concept. In one plant, the operator is on one side of an Aaronson workpiece positioner, and the manipulator is on the other. The operator sets up a second workpiece while the manipulator works on the first. When both are finished with their tasks, the positioner rotates to swap the positions of the two workpieces. The next workpiece is then said to have docked at its processing station. While the manipulator processes the new workpiece, the person removes the previous one and replaces it with another new one. This system protects the worker in two ways: (1) the worker never comes within reach of the manipulator and (2) the bulk of the positioner lies between the worker and the manipulator in case it should throw something. The positioner is fastened securely to the floor so that the manipulator cannot push it over onto the worker.

Appropriate workplace design can reduce the amount of damage done by impacts when they do occur. A rigid, inexpensive work surface which will give under an impact can be built from a layer of styrofoam several inches thick covered by plywood. Hand tooling and bench fixtures should have shear pins, ball detents, or preload springs at strategic places to permit them to yield or break away in response to excessive forces and sensors to detect when that happens.

A hydraulic arm should never be operated with its covers off; if a hydraulic line ruptures, combustible oil will spray all over.

6.3.7 Restricting Arm Motion

When people must work within the arm's reach, it should be constrained to move slowly. One way to restrict the speed of a manipulator with hydraulic actuators is to place a flow restriction in the fluid line which runs to the servo valves from the accumulator (or from the pump if there is no accumulator). This flow restriction will permit the arm to lift its rated load but will prevent it from moving at full speed. Corresponding methods for slowing down an electric arm are less reliable and more complex. In order to restrict the force which a hydraulic arm can exert, one can bypass each actuator through a suitable overpressure relief valve. However, some actuators must still be able to overcome considerable gravity loads. The suggestion that the robot be restricted in movement with the installation of steel posts in the floor is rejected, as it would be better to be pushed over than to be pinned to a post or any other restrictive device (Reference 83).

Sensors may be used to protect the arm from collisions with objects. One approach is to place light-beam barriers around the normal working volume for the task to be performed. If the arm interrupts the light beam, the arm will be shut down. Another approach is to mount sensors on the arm itself. For example, one might mount one or more light-beam barriers parallel to each link of the arm. Various kinds of proximity detectors such as microswitches operated by cat-whisker feelers, infrared proximity detectors, and ultrasonic ranging devices, could also be mounted on the arm. Whatever type of sensor is chosen, it must be rugged enough and must reliably detect the presence of arbitrary objects.

6.3.8 Operator Training

Accidents cannot be prevented by safety devices alone. Those who work with or around robots must also be trained in the precautions necessary for their own safety. For example, it is educational to see a robot snap a 3/8-inch steel rod in two.

Some of the mistakes that people commonly make and that a training program should aim to eliminate are the following:

1. If the arm is not moving, they assume it is not going to move.
2. If the arm is repeating one pattern of motions, they assume it will continue to repeat that pattern.
3. If the arm is moving slowly, they assume it will continue to move slowly.
4. If they tell the arm to move, they assume it will move the way they want it to.

In summary, use good common sense in all aspects of the application and check each part of the engineering for safe practices as for any other piece of automated equipment.

6.3.9 OSHA Regulations*

In order to help ensure that particular manufacturing operations are free from recognized hazards to workers, industrial robots are being used and considered for a number of jobs covered by OSHA standards. Various Subparts of Part 1910, Occupational Safety and Health Standards, extend somewhat indirectly into the use of industrial robots. Chief among these are

Subpart G - Occupational Health and Environmental Control

Subpart H - Hazardous Materials

* The material in this section is adapted from Heroux, N. M. and G. Munson, Jr., "Robots Reduce Exposure To Some Industrial Hazards," Industrial Robots, Fundamentals, Volume 1, SME, 1979.

Subpart I - Personal Protective Equipment

Subpart L - Fire Protection

Subpart N - Materials Handling and Storage

Subpart O - Machinery and Machine Guarding

Subpart P - Hand and Portable Powered Tools and Other Hand-Held Equipment

Subpart Q - Welding, Cutting, and Brazing

Subpart S - Electrical

The listing cited is not to be considered complete. In addition, some of the relationships between the OSHA regulations and the use of robots may be quite remote with clarification depending upon the specific situation under consideration. At present, robots are not involved in every one of the cited job areas, but their capabilities are such that they could serve in some capacity.

As with all OSHA standards, such applications are determined basically by the type of industry or nature of the operation involved. Hence, the industral user of robots should be familiar with all of the many OSHA references that pertain to his plant operation. Even though robots may be used, OSHA regulations still must be complied with in order to protect those employees entering robot station areas and to safeguard employees working in areas adjacent to where the robots are located and where such hazards as noxious fumes, excessive noise, or extreme heat may be present.

Specific questions should be referred to OSHA compliance officers or to the OSH-Administration area director.

6.4 JUSTIFICATION FOR THE USE OF ROBOTICS

In most cases, the major factors justifying the use of robotic technology for manufacturing are economic in nature. An industrial robot manufacturing system represents a sizeable capital investment and hopefully an even more sizeable return-on-investment. Noneconomic factors provide certain intangible benefits that may justify using robotic technology. Some of these factors are increased productivity, improved quality and utilization of materials, performance of hazardous operations and undesirable tasks, advancement of manufacturing technology, adaptability, competitive advantage, and management direction.

6.4.1 Noneconomic Factors

In most cases, increased productivity results from the robotic ability to maintain a constant pace throughout the entire work shift, rather than the robotic ability to perform tasks faster than a person. In some cases, a person can outperform a robot in the speed with which a task is completed, especially in some complex manipulative tasks. However, a person usually cannot maintain this performance level for an entire shift because of fatigue. Generally, robotic technology provides a tireless worker and increased productivity by maintaining a constant rate of production for extended periods

often

of time. The average cycle time for parts tends to be lower for robots. The result is that more parts are produced per shift, and this increased productivity represents an economic gain.

These gains are realized through the consistent operation of the robot. Once an optimum procedure is defined and programmed into the robot for a particular task, that task will be consistently performed in the optimum manner every time. The result is fewer bad parts that have to be scrapped, less material waste, and measurable economic gains.

Robots are frequently used to perform operations that are potentially hazardous to human workers, usually in order to comply with safety regulations. These hazardous operations include press loading and unloading and working in toxic atmospheres or extremes of ambient temperatures. Improved safety can result in reduced operating costs and provide some economic justification. However, a robotic solution to safety problems may not always be economically desirable, and wisdom dictates that alternatives also be investigated.

Robots can be used to perform some undesirable tasks, and the result is additional costs savings. If ignored, workers' complaints about poor working conditions, i.e., excessive noise, dust, fumes, heat, dirt, heavy loads, fast pace, or monotony, can lead to work stoppages or slowdown, uncompleted operations, poor workmanship, high labor turnover, absenteeism, grievances, or sabotage, and can result in higher-than-normal operating costs. Compensation may require overtime to make up production losses, rework and repair, and expenses for processing grievances, hiring replacement personnel, and training new workers. In many cases, robotics is a cost-effective solution.

A robot may be introduced for the advancement of technology. In such applications, one or a limited number of robots may be installed for developmental purposes. The intent here is to gain the knowledge and expertise required to implement similar robotic applications in an actual production setting where economic benefits are more direct. Economic returns usually are realized in the follow-on production applications. In fact, the costs of a developmental implementation are often factored into the cost of the follow-on production application.

Industrial robots are adaptable. Whether programmable or not, most have a degree of adaptability that allows them to be moved around or used in different types of tasks, thereby increasing their usefulness and potential return on investment. Few managers will accept, however, the adaptability of a robot as a justification, without a plan for how it can and will be used in different applications. Adaptability may be a possibility but should not be used as a justification factor unless a serious implementation plan is established. Usually the feasibility of adapting to other applications will diminish as development progresses due to the enhancements specifically designed for the primary application (i.e., tooling, facilities layout). Adaptability is an intangible asset that may become useful in a contingency situation.

Enhancement of competitive position has both direct and indirect economic implications. Direct benefits are obvious. Lower production costs resulting from the use of robots give a manufacturer a pricing advantage over his competitors. The inherent flexibility of robot manufacturing systems indirectly affects the economics. Shifting market demands are easily met by increasing or decreasing production rates on various products without changing the size of the work force. New products can also be introduced quickly and easily, often with little change to production facilities.

Occasionally, a robotic implementation may be made on the basis of management direction. This alone is a poor justification for implementation, especially if the determination is not based on economic considerations. Projects based solely on this type of impetus tend to have a low success rate. Efforts to comply with the directive may result in careless choice of robot or application. The application may become more complex than first anticipated, or the robot may not have the necessary capabilities to perform the tasks. Aside from the obvious waste of capital, a bad experience may discourage management from further attempts to implement robotic technology, even when other potentially successful applications may exist.

Although the previously discussed factors should play a key role in the evaluation of a robotics installation, the weight of the final decision should rest on a firm economic foundation. Economic considerations fall into two major categories - cost avoidance and cost savings.

6.4.2 Economic Analysis

There are numerous methods of economic analysis for any capital investment. The selection of a method depends on the size of the investment, the amount of risk involved, the projected life of the investment, company financial condition, whether or not the investment is for new or replacement equipment, management policy, and many other criteria determined by the situation.

An economic analysis is basically a systematic examination of a complex business activity that will aid in making a decision about a capital investment by providing a basis upon which to make a decision. If the analysis is undertaken to justify a decision already made, the true purpose of the analysis is misguided.

In general, there are two situations for which an economic analysis is used. The first situation involves investment in equipment for a new application or to avoid costs; the second involves an investment to replace an existing method.

In the first case, the purpose of the analysis is to identify the least expensive method with which to accomplish a task. The second case is a comparison between the present method and one or more new methods. The task of justification in the second case becomes difficult because it is to be based on investment cost compared to savings over the cost of an existing method. Since the savings are determined relative to the present method, there is no absolute measure of profitability because the savings depend as much on how bad the present method is, as on how good the proposed method is.

The life cycle of a capital investment will typically follow a pattern as shown in Figure 30. Initially, money flows out until the project comes on line. From then on, savings first recover the investment and then produce net earnings. The project first breaks even and later recovers all of the earlier negative cash flows to produce net earnings.

6.4.2.1 Batch Manufacturing Contingencies

Several factors have been selected that should be considered in an investment analysis in aerospace batch manufacturing. In this type of environment, production rates are established according to the number of units (aircraft, etc.) contracted for during a time period and are limited to facilitate design changes during the life of the contract. Production volume then is established by the batch lot sizes required to meet the unit rates; therefore, machine capability over the batch requirement is of little value.

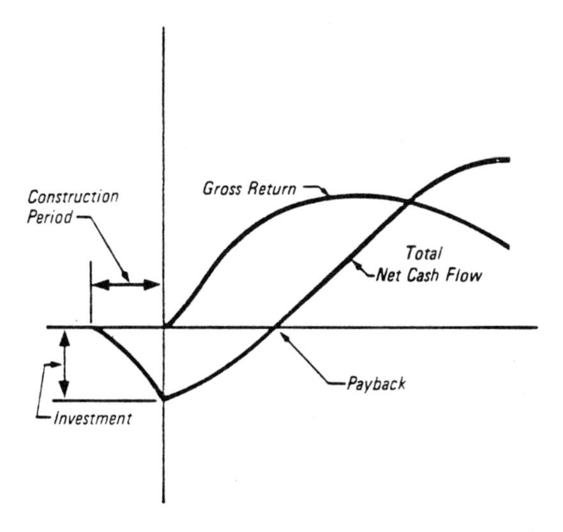

Figure 30
LIFE CYCLE OF A CAPITAL INVESTMENT

There are exceptional cases, however, due to the unpredictability of the market, where volume capability is useful. A potential high production capability to support large volume manufacturing, as may be necessary for national defense, would be advantageous and desirable.

Product design restrictions may place certain limitations on the available investment options. For example, a robot capable of efficiently installing pop-rivets would not be a feasible investment for a method of aircraft structural assembly because the design probably specifies a more reliable fastener.

The attitude toward investment in new technology in the aerospace industry is very good. Competition within the industry and the potential for profit, derived from better methods of producing the products of the future, is responsible for this attitude. It is therefore important to provide accurate analyses of capital investments for decisions which will, in all likelihood, have long-term effects.

6.4.2.2 Analysis Methods

In this discussion, we will examine methods for analyzing investments in terms of their savings over present methods (as in the second case mentioned above). Capital investment analysis for new business prospects not related to current applications will be left for marketing and financial analysts to provide. The more familiar analysis methods to be discussed are return on investment (ROI), break-even or payback analysis, and discounted cash flow.

There is no single method for determining return on investment (ROI). It is generally used to compare a prospective machine's savings to its investment. The savings are divided by the investment to get a rate of return (percentage). ROI can be calculated for first-year or full-life return. On large investments, such as robots, it is best to use full-life return. This provides a method that considers the total benefits over the life of the investment. A more accurate analysis can be made by considering the time-value of money, as in discounted cash flow analysis (explained later).

Payback is the length of time it takes to recover investment costs. It is found by determining how long it will be until the project's gross after-tax cash flow equals its investment. Company policy will determine what payback period is acceptable. In general, a robot can be expected to reach payback in one to four years. The payback period is shown in Figure 30.

The idea of discounted cash flow is simple. All future costs and all expected income for the life of the investment are converted to their present value (the value of future net-cash-flow today) and compared. Discounting allows everything to be put on a current-dollar basis and puts the investment into perspective with other investment opportunities. In other words, it compares the capital investment to a financial investment that will return a certain rate of return, usually 25 or 30 percent. Although the discounted cash flow idea is simple, its calculation can be complicated and, as stated before, project savings are only savings as they compare to a current method.

For a robotic investment, the three methods presented can be used in a combined analysis. In this way, return on investment, payback, and discounted cash flow are maximized for a realistic analysis. To begin the analysis, information about present methods, production rates, and savings factors must be compiled.

6.4.2.3 Data For Analysis

The data that is needed can be divided into two categories, investments data and savings data, corresponding to the first two divisions on a typical investment form as shown in Figure 31. Table 1 provides a description of each item in the investment form. Under investments, the entire cost for planning, developing, purchasing, and installing the robot and equipment for the project is listed. In our example, all costs are incurred during the first year.

The operating savings distribution (Figure 32) is used to calculate the dollar savings for the operating savings/cost section. For this example, the robot will be utilized on only one shift. Labor savings are calculated from the net costs between the current and proposed methods. For this example, the current method is manual and the robot is expected to increase production 2.8 times. Therefore, with the proposed method there is a net labor savings. The remaining savings calculations are made by comparing the current costs and proposed method costs to arrive at the net savings. In the example, indirect labor, maintenance, and other costs are negative, indicating the proposed method has higher costs in these areas.

PROJECT TITLE: ROBOT ECONOMIC ANALYSIS EXAMPLE

		YEAR								TOTAL
		1	2	3	4	5	6	7	8	
INVESTMENT										
CAPITAL FACILIITES										
1. Robot Cost		$ 65,000								$ 65,000
2. Accessories Cost		5,000								5,000
3. Related Expense		20,000								20,000
DEVELOPMENT										
4. Engineering		170,000								170,000
5. Installation		5,000								5,000
6. Tooling		15,000								15,000
7. **TOTAL**		280,000	0							280,000
OPERATING SAVINGS/COSTS										
LABOR SAVINGS										
8. Direct		90,000	90,000							720,000
9. Indirect		(4,000)	(4,000)							(32,000)
10. Maintenance Savings		(2,300)	(2,300)							(18,400)
11. Other Savings		7,000	7,000							56,000
12. Other Costs		(7,500)	(2,500)							(25,000)
13. **TOTAL**		83,200	88,200							700,600
ANALYSIS										
14. Total Investment	Line 7	280,000	0							280,000
15. Total Savings	Line 13	83,200	88,200							700,600
16. Net Depreciation		35,000	35,000							280,000
17. Net Savings	Line 15 – Line 16	48,200	53,200							420,600
18. Tax Rate 48%										
19. Net Savings After Tax	Line 17 x .52	25,064	27,664							218,712
20. Total Cash Return	Line 19 + Line 16	60,064	62,664							498,712
21. Net Cash Flow	Line 20 – Line 14	(219,936)	62,664							218,712
22. Discount Factor 20%		.833	.694	.578	.482	.402	.335	.279	.233	xxx
23. Discounted Cash Flow	Line 21 x Line 22	(183,207)	43,488	36,219	30,204	25,190	20,992	17,483	14,600	4,973
24. Discount Factor 25%		.800	.640	.512	.410	.328	.262	.210	.168	xxx
25. Discounted Cash Flow	Line 21 x Line 24	(175,949)	40,105	32,084	25,692	20,553	16,418	13,159	10,527	(17,409)

FIGURES FOR YEARS 2 THROUGH 8 ARE IDENTICAL

26. **Return on Investment 21.1%**
27. **Payback 4.5 years**

4,973 – 22,382 x 5 + 20.00 = 21.11%

Figure 31
ECONOMIC ANALYSIS FORM

Data for the analysis section is now compiled. The investment and savings figures are entered and the depreciation schedule is calculated. Eight-year straight-line depreciation and a tax rate of 48 percent was used in calculating the net savings after tax. Total cash return is calculated by taking the project savings minus tax plus the depreciation and is used in calculating the payback period for the project. The net cash flow is determined for use in calculating the discounted cash flow and ROI.

The discount factors are taken from a discount factors table. Each year's net cash flow is multiplied by the discount factor and then the yearly discounted values are totaled. Generally, using this method, the ROI is calculated by interpolating between discount factors, as shown in the example. The payback period is the time required for total cash return to equal the investment. If the yearly cash flow is constant, the investment can be divided by this constant for calculation of the payback period in years.

The resulting analysis for the example shown in Figure 31 for a single shift application indicates a return on investment of 21.1 percent and a payback period of 4.5 years. To provide data for comparison, a similar analysis, using similar investment data and utilizing the robot on two shifts, was compiled and the results are indicated below.

Application	Investment	Savings	ROI	Payback
One Shift	$280,000	$ 700,600	21.1%	4.5 years
Two Shift	$380,000	$1,401,200	34.1%	3.3 years

As expected, two-shift utilization yields a greater return and provides an earlier payback even with a substantial increase in investment. During the economic analysis process, the adjustment of various factors (such as utilization) and the comparison of each option is beneficial not only in justifying the investment, but in planning for an optimum return as well. It can be noted here that the analysis may or may not include some factors which could affect the feasibility of the investment. For example, increasing labor costs or declining productivity rates, which may influence the investment decision, may not be projected in the analysis.

Although the use of robots may be justified for a variety of reasons, most motivation will be supplied by the economics of the situation. A successful justification requires consideration and quantification of all potential costs and cost benefits other than direct-labor replacement factors. Many of the cost factors can only be estimated during the justification preparation. However, following the installation of the robot, its actual cost performance usually can be easily and accurately measured. It is important, therefore, that the original estimates be as accurate as possible.

Table 1 ECONOMIC ANALYSIS ITEMS

ITEM	DESCRIPTION
1. Robot Cost	Basic cost of the robot, operational equipment, maintenance and test equipment included in the basic price of the robot.
2. Accessories Costs	Additional equipment, optional and required, that is purchased for the robot (includes additional hardware, recorders, testers, computers and tools).
3. Related Expense	Should include all additional hardware costs and expenses for the application (such as conveyors, guard rails, component cabinets, interface hardware, insurance, etc.).
4. Engineering Costs	Estimated cost of planning and design in support of project development (includes research and laboratory expense).
5. Installation Costs	Labor and materials for site preparation, floor or foundation work, utility drops (air, water, electricity), and set-up costs.
6. Tooling Costs	Labor and materials for special tooling (end-effectors), interface devices between controller and tooling, fabrication of part positioners, fixtures, and tool controllers.
7. Total Cost of Investment	Sum lines 1 through 6.
8. Direct Labor Savings	Net direct labor savings realized from converting to the proposed method (compares costs of direct labor, benefits, allowances, shift premiums, etc., and may include overhead costs to simplify calculations).
9. Indirect Labor Savings	Net indirect labor savings realized from converting to the proposed method (maintenance, repair, and other related labor support costs).
10. Maintenance Savings	Estimate of net maintenance savings to be realized from conversion to the proposed method (includes maintenance supplies, replacement parts, spare parts, lubricants, service contract charges, etc.).

Table 1 ECONOMIC ANALYSIS ITEMS

ITEM	DESCRIPTION
11. Other Costs	Increased or additional costs of the proposed method over the current method for supplies, utilities, training, etc. (Initial training costs are included; however, ongoing training is not, since it is assumed that is will not change the net ongoing cost.)
12. Other Savings	Savings or cost reductions of the proposed method compared to the current method. (includes material savings, i.e., reduced scrap; and reduced downtime, i.e., nonproductive time, etc.).
13. Total Operating Savings	Sum lines 8 through 12.
14. Total Investment	From line 7.
15. Total Savings	From line 13.
16. Depreciation	Yearly depreciation calculated using straight line, declining balance, or some other method (line 14 multiplied by the yearly percentage).
17. Net Savings Before Tax	Total savings minus depreciation (line 15 minus line 16).
18. Tax Rate	Corporate tax rate, approximately 52 percent.
19. Net Savings After Tax	Line 17 multiplied by line 18.
20. Total Cash Return	Line 19 plus line 16.
21. Net Cash Flow	Line 20 minus line 14.
22. Discount Factor	For calculation of present value.
23. Discounted Cast Flow	Line 21 multiplied by line 22.
24. Discount Factor	For calculation of present value.
25. Discounted Cash Flow	Line 21 multiplied by line 24.

Table 1 ECONOMIC ANALYSIS ITEMS

ITEM	DESCRIPTION
26. ROI	Percent return on investment, discounted and interpolated between the lower and upper discount factors (calculated by dividing the total on line 23 by the difference between the totals on lines 23 and 25, then multiplying this by the number of years of project life and adding the total to line 22).
27. Payback Period	Time period from the start of the project until line 20 exceeds line 14.

OPERATING SAVINGS/COST ANALYSIS

Labor			
Direct			
Current Method	$ 140,000		
Proposed Method	$ 50,000		
Net Direct Labor Savings			$ 90,000
Indirect			
Current Method	$ 500		
Proposed Method	$ 4,500		
Net Indirect Labor Savings			$ (4,000)
Maintenance			
Current Method	$ 1,680		
Proposed Method	$ 3,980		
Net Maintenance Savings			$ (2,300)
Other Savings			
Reduced Scrap	$ 7,000		
Total Other Savings			$ 7,000
Other Costs	1st yr.	after 1st yr.	
Training	$ (5,000)	0	
Supplies	$ (2,500)	(2500)	
			$ (7,500) | $ (2,500)
Net Total Operating Savings			$ 83,200 | $ 88,200

Figure 32
SAVINGS DISTRIBUTION FORM

References

1. H. Makino, "A Kinematical Classification of Robot Manipulator," in Proc. 3rd Conf. on Indust. Robot Technol., Tokyo, Japan, Nov. 1977 (Bedford, England: Int'l. Fluidics Services, Ltd.) and Proc. 6th Int'l Symp. on Indust. Robots, Nottingham, England, March 1976.

2. A. d'Auria and M. Salmon, "SIGMA - An Integrated General-Purpose System for Automatic Manipulation," Proc. 5th Int'l Symp. on Indust. Robots, Chicago, IL, 22-24 September, 1975 (Chicago, IL: IIT Research Inst.), pp. 185-202.

3. "Pneumatic Servomechanisms Dynamic Analysis Program," Tech. Supp. Package MFS-23295 (Wash. D.C.: Nat'l Aeron. & Space Admin., Technol. Util. Branch), summary in NASA Tech Briefs (Spring 1978), p. 146.

4. P. J. Drazen and M. F. Jeffery, "Some Aspects of an Electro-Pneumatic Industrial Manipulator," Proc. 8th Int'l Symp. on Indust. Robots, Vol. I, Stuttgart, 30 May - 1 June, 1978 (Bedford, England: Int'l. Fluidics Services, Ltd.), pp. 396-405.

5. R. L. Paul, "Robot Software and Servoing," Proc. NSF Workshop on the Impact on the Academic Community of Required Research Activity for Generalized Robotic Manipulators, Gainesville, FL, 8-10 Feb., 1978 (Gainesville, FL: M.E. Dept., U. Florida), pp. 255-259.

6. M. Blanchard, "Digital Control of a Six-Axis Manipulator," Working Paper 129 (Cambridge, MA: MIT Artif. Intel. Lab., 1976).

7. M. Raibert, "A State Space Model for Sensorimotor Control and Learning," Memo 351, Nat'l Inst. of Health Grant NIH-5-T01-6MO-1064-14, Office of Naval Res. Contract N00014-75-C-0634 (Cambridge, MA: MIT Artif. Intel. Lab., Jan. 1976).

8. W. Snyder, "Computer Control of Robots--A Servo Survey," MS76-617 (Dearborn, MI: Soc. Manuf. Eng.'s, Mkt. Svc. Dept., 1976).

9. R. L. Paul, et al., "Advanced Industrial Robot Control Systems," 1st Report, N.S.F. Grant APR-77-14533, Report No. TR-EE 78-25 (West Lafayette, IN: Purdue U., School of E. E., May 1978).

10. R. L. Paul, et al., "Advanced Industrial Robot Control Systems," 2nd Report, N.S.F. Grant APR-77-14533, Report No. TR-EE 79-35 (West Lafayette, IN: Purdue U., School of E. E., July 1979).

11. B. L. Davies and E. Ihnatowicz, "A Three-Degree-of-Freedom Robotic Manipulator," in Proc. Robots IV Conf., Detroit, MI, 30 Oct. - 1 Nov. 1979.

12. B. L. Davies and E. Ihnatowicz, "A Three-Degree-of-Freedom Robotic Manipulator," Robotics Today (Winter 1979-80), pp. 28-29.

13. J. S. Albus, "A New Approach to Manipulator Control: The Cerebellar Model Articulation Controller (CMAC)," J. Dynamic Systems, Measurement and Control, Trans. of the A.S.M.E. (Sep. 1975), pp. 220-227.

14. J. S. Albus, "Data Storage in the Cerebellar Model Articulation Controller (CMAC)," J. Dynamic Systems, Measurement and Control, Trans. of the A.S.M.E. (Sep. 1975), pp. 228-233.

15. J. J. Craig and M. H. Raibert, "A systematic Method of Hybrid Position/Force Control of a Manipulator," Proc. Computer Software & Applications Conf., Chicago Il., 6-8 Nov. 1979, pp. 446-451.

16. B. Dobrotin and R. A. Lewis, "A Practical Manipulator System," 5th Int'l Jt. Conf. on Artificial Intelligence, Cambridge, MA, 22-25 Aug., 1977 (Pittsburgh, PA: Carnegie-Mellon U., Dept. of Comp. Sci.), pp. 723-732.

17. J. Y. S. Luh and C. S. Lin, "Multiprocessor-Controllers for Mechanical Manipulators," Proc. Computer Software & Applications Conf., Chicago, IL., 6-8 Nov. 1979, pp. 458-463.

18. K. Young, "Controller Design for a Manipulator Using Theory of Variable Structure Systems," I.E.E.E. Transactions on Systems, Man, and Cybernetics, Vol. SMC-8, No. 2 (Feb. 1978), pp. 101-109.

19. M. J. Dunne, "An Advanced Assembly Robot," (Unimate 6000 series dual manipulator) Paper MS77-755 (Dearborn, MI: Soc. Manuf. Eng.'s, Mkt. Svc. Dept., 1977).

20. V. Utkin, "Equations of Sliding Mode in Discontinuous Systems" Automation and Remote Control, Part I (1971), pp. 1897-1907, and Part II (1972), pp. 211-219.

21. R. L. Paul, "Modelling, Trajectory Calculation and Servoing of a Computer Controlled Arm," Memo AIM-177, Report STAN-CS-72-311 (Stanford, CA: Stanford Artif. Intel. Proj., Nov., 1972).

22. D. Peiper, "The Kinematics of Manipulators under Computer Control," Ph. D. Thesis, A.R.P.A. Order No. 457, Contract SD-183, Report CS-116 or Memo AI-72 (Stanford, CA: Stanford Artif. Intel. Proj., Oct. 1968).

23. C. A. Rosen, D. Nitzan, et al., "Exploratory Research in Advanced

Automation," 5th Report, NSF Grant GI38100X1, SRI Project 4391, (Menlo Park, CA: SRI Int'l, Artif. Intel. Ctr., Jan. 1976).

24. D. Whitney, "Resolved Motion Rate Control of Manipulators and Human Prostheses," I.E.E.E. Trans. Man-Machine Systems, Vol. MMS-10, No. 2 (June 1969), pp. 47-53.

25. T. Durham, "Infrared Light for a New Wireless Revolution," New Scientist, Vol. 84 (double issue) No. 1186/1187, Dec. 20/27, 1979, pp. 931-933.

26. "Non-contacting Ultrasonic Distance Measurement," (U.K. Patent Application 803/75) NRDC Inventions (London, England: Nat'l Res. & Devel. Corp., July 1976).

27. "Sensors Give Eyes New Horizons," New Scientist, Vol. 84, No. 1185 (Dec. 1979), p. 675.

28. D. Nitzan, A. E. Brain and R. O. Duda, "The Measurement and Use of Registered Reflectance and Range Data in Scene Analysis," Nat'l Sci. Found. Grant ENG-75-09327, and Advanced Res. Proj's Agency Contract DAHCO4-72-C-0008, Proc. IEEE, Vol. 65, No. 2 (Feb. 1977), pp. 206-220.

29. A. K. Bejczy, "Smart Sensors for Smart Hands," in Remote Sensing of Earth from Space -- Role of "Smart Sensors," R. A. Breckenridge, (ed.), Vol. 67 of Progress in Astronautics and Aeronautics, M. Sommerfield, (ed. of series) (New York, NY: Amer. Inst. of Aero. and Astro., 1979), pp.275-304.

30. T. O. Binford, et al., "Exploratory Study of Computer Integrated Assembly Systems, Progress Report 4," Memo AIM-285.4 and Report STAN-CS-76-568 (Stanford, CA: Stanford Artif. Intel. Proj., June 1977), 250 pp.

31. P. Watson and S. Drake, "Pedestal and Wrist Force Sensors for Automatic Assembly," Report P-176 (Cambridge, MA: C. S. Draper Labs, Inc., June 1975), and in Proc. 5th Int'l Symp. on Indust. Robots, Chicago, IL, 22-24 September, 1975 (Chicago, IL: IIT Research Inst.), pp. 501-512.

32. B. E. Shimano, "The Kinematic Design and Force Control of Computer-Controlled Manipulators," Ph. D. Thesis, Stanford U. Mech. Eng. Dept., Memo AIM-313 and Report STAN-CS-78-660, supported by Advanced Res. Proj's Agency Contract MDA903-76-C-0206 and the Nat'l Sci. Found., (Stanford, CA: Stanford Artif. Intel. Proj., March 1978), 134 pp.

33. G. Agin, "Vision Systems for Inspection and Manipulator Control," Proc. 1977 Jt. Automatic. Control Conf., San Francisco, CA, 22-24 June, 1977 (Piscataway, NJ: IEEE Svc. Ctr.), pp. 132-138.

34. G. Agin, "Real Time Control of a Robot with a Mobile Camera," Tech. Note 179 (Menlo Park, CA: SRI Int'l, Artif. Intel. Ctr., Feb. 1979).

35. G. Gleason, and G. Agin, "A Modular Vision System for Sensor-Controlled Manipulation and Inspection," Proc. 9th Int'l Symp. on Indust. Robots, Wash., D.C., 13-15 March, 1979 (Dearborn, MI: Soc. Manuf. Eng.'s, Mkt. Svc. Dept.), pp. 57-70, and Tech. Note 178 (Menlo Park, CA: SRI Int'l, Artif. Intel. Ctr., Feb. 1979).

36. "Camera Zooms in on Auto Focusing," New Scientist, Vol. 83, n. no. (June 8, 1978), p. 672.

37. G. J. VanderBrug, J. S. Albus, and E. Barkmeyer, "A Vision System for Real Time Control of Robots," Proc. 9th Int'l Symp. on Indust. Robots, Wash., D.C., 13-15 March, 1979 (Dearborn, MI: Soc. Manuf. Eng.'s, Mkt. Svc. Dept.), pp. 213-232.

38. G. J. Vanderbrug, J. S. Albus, and E. Barkmeyer, "A Vision System for Real-Time Robot Control," Robotics Today (Winter 1979-80), pp. 20-22.

39. A. P. Ambler, et al., "A Versatile Computer-Controlled Assembly System," 3RD Int'l Jt. Conf. on Artificial Intelligence, Stanford, CA, 20-23 Aug. 1973 (Menlo Park, CA: SRI Int'l, Artif. Intel. Ctr.) †, pp. 298-307.

40. W. P. Buttler, "Self-Navigating Robot," Tech. Supp. Package NPO-14190 (Wash. D.C.: Nat'l Aeron. & Space Admin., Technol. Util. Branch), summary in NASA Tech Briefs (Spring 1978), pp. 31-32.

41. G. Agin, "Servoing with Visual Feedback," Technical Note 149 (Menlo Park, CA: SRI Int'l, Artif. Intel. Ctr., July 1977).

42. Prajoux, R., "A Step Toward the Handling of Parts Carried by an Overhead Conveyor: A Robot System using a Fast Vision Sensor to Track a Hanging Object," supported by NSF Grant APR-75-13074, and by C.N.R.S and I.R.I.A. (France) Tech. Note 208 (Menlo Park, CA: SRI Int'l, Artif. Intel. Ctr., Dec. 1979).

43. S. Kashioka, et al., "An Approach to the Integrated Intelligent Robot with Multiple Sensory Feedback: Visual Recognition Techniques," Proc. 7th Int'l Symp. on Indust. Robots, Tokyo, 19-21 October, 1977 (Tokyo: Japan Indust. Robot Assoc.), pp. 531-538.

44. S. Hirose and Y. Umetani, "The Development of Soft Gripper for the Versatile Robot Hand," Proc. 7th Int'l Symp. on Indust. Robots, Tokyo, 19-21 October, 1977 (Tokyo: Japan Indust. Robot Assoc.), pp. 353-360.

45. A. Rovetta and G. Casarico, "On the Prehension of a Robot

Mechanical Hand: Theoretical Analysis and Experimental Tests," Proc. 8th Int'l Symp. on Indust. Robots, Vol. I, Stuttgart, 30 May - 1 June, 1978 (Bedford, England: Int'l. Fluidics Services, Ltd.), pp. 444-451.

46. J. E. Griffith, et al., "Quasi-Liquid Vise for a Computer-Controlled Manipulator," Report RC 5451, . 23797 (Yorktown Heights, NY: IBM T. J. Watson Res. Ctr., June 1975).

47. F. Skinner, "Design of a Multiple Prehension Manipulator System," ASME Publ. No. 74-DET-25 (Oct. 1974).

48. F. Skinner, "Multiple Prehension Hands for Assembly Robots," Proc. 5th Int'l Symp. on Indust. Robots, Chicago, IL, 22-24 September, 1975 (Chicago, IL: IIT Research Inst.), PP. 77-88.

49. J. G. Bollinger, and P. W. Ramsey, "Computer Controlled Self Programming Welding Machine," in Welding Journal (May, 1979).

50. R. Mosher, "Robotic Painting -- the Automotive Potential," Paper MS77-735 (Dearborn, MI: Soc. Manuf. Eng.'s, Mkt. Svc. Dept., 1977).

51. T. Kuzmierski, "Robot Development for Aerospace Batch Manufacturing," Proc. 1977 Jt. Automatic. Control Conf., San Francisco, CA, 22-24 June, 1977 (Piscataway, NJ: IEEE Svc. Ctr.), pp. 704-709.

52. J. H. Lockett, "Small-Batch Production of Aircraft Access Doors Using an Industrial Robot," in Proc. Robots IV Conf., Detroit, MI, 30 Oct. - 1 Nov. 1979.

53. J. H. Lockett, "The Robotic Work Station in Small-Batch Production," Robotics Today (Winter 1979-80), pp. 17-19.

54. "Rigid Coupling is Also Flexible," Tech. Supp. Package MSC-16488 (Wash. D.C.: Nat'l Aeron. & Space Admin., Technol. Util. Branch), summary in NASA Tech Briefs (Spring 1978), p. 105.

55. D. McGhie, and J. Hill, "Vision-Controlled Subassembly Station," Paper MS78-685 (Dearborn, MI: Soc. Manuf. Eng.'s, Mkt. Svc. Dept., Nov. 1978), and in Proc. Robots III Conf., Chicago, IL, 7-9 Nov. 1978, supported by Nat'l Sci. Found. Grant APR75-13074, reprints available (Menlo Park, CA: SRI Int'l, Artif. Intel. Ctr.)

56. "Tool Wear Sensors," 4th NSF/RANN Grantees' Conference on Production Research and Technology, Chicago, IL, 30 Nov. - 2 Dec., 1976 (Wash., D.C.: Nat'l Sci. Found., Div'n Advanced Productivity Res. & Technol.), pp. 68-73.

57. J. W. Hill and Sword A. J., "Programmable Part Presenter Based on Computer Vision and Controlled Tumbling," supported by Nat'l Sci.

Found. Grants APR-75-13074 and 15 industrial affiliates, SRI Projects 4391 and 6284, Tech. Note 194 (Menlo Park, CA: SRI Int'l, Artif. Intel. Ctr., March 1980), 12 pp.

58. G. E. Ennis and M. A. Eastwood, "Robotic System for Aerospace Batch Manufacturing," 2nd Quarterly Interim Technical Report, 12 Nov. - 12 Feb., 1979, U.S.A.F. Contract F33615-78-C-5189, Task B (Wright-Patterson Air Force Base, OH: AFML/LTC ICAM Proj., Feb. 1979), pp. 2.23-2.31.

59. J. A. Feldman, "Programming Languages," Scientific American, Vol. 241, No. 6 (Dec. 1979), pp. 94-112.

60. "Preliminary ADA Reference Manual," ACM SIGPLAN Notices, Vol. 14, No. 6, Part A (June 1979).

61. "Rational for the Design of the ADA Programming Language," ACM SIGPLAN Notices, Vol. 14, No. 6, Part B (June 1979).

62. T. Lozano-Perez, "The Design of a Mechanical Assembly System," Master's Thesis, Report No. 397 (Cambridge, MA: MIT Artif. Intel. Lab., 1978).

63. R. H. Taylor, "A Synthesis of Manipulation Control Programs from Task-Level Specifications," Ph. D. thesis, Stanford U., Memo 282 (Stanford, CA: Stanford Artif. Intel. Proj., 1976).

64. L. I. Lieberman and M. A. Wesley, "AUTOPASS: A Very High Level Programming Language for Mechanical Assembler Systems," Report RC 5599 (#24205) (Yorktown Heights, NY: IBM T. J. Watson Res. Ctr., Aug. 1975).

65. "User's Guide to VAL," (Danbury, CT: Unimation Inc., Feb. 1979).

66. R. L. Paul, "WAVE - A Model-based Language for Manipulator Control," The Industrial Robot, Vol 4, No. 1 (March, 1977), pp. 10-17.

67. S. Mujtaba and R. Goldman, "AL Users' Manual," Memo AIM-323 and Report STAN-CS-79-718, Nat'l Sci. Found. Grants NSF-APR-74-01390-A04 and NSF-DAR-78-15914, (Stanford, CA: Stanford Artif. Intel. Proj., Jan. 1979), 130 pp.

68. L. I. Lieberman and M. A. Wesley, "AUTOPASS: An Automatic Programming System for Computer-Controlled Mechanical Assembly," IBM J. Res. & Devel., Vol. 21 (July 1977), pp. 321-333.

69. W. T. Park, "Minicomputer Software Organization for Control of Industrial Robots," Proc. 1977 Jt. Automatic. Control Conf., San Francisco, CA, 22-24 June, 1977 (Piscataway, NJ: IEEE Svc. Ctr.), pp. 164-171.

70. "When Machines Talk and Listen," Design Engineering Feature, in Product Engineering, October, 1978.

71. S. Tsuruta, "DP-100 Voice Recognition System Achieves High Efficiency," in JEE, July, 1978 (Tokyo, Japan: DEMPA Publications), and reprint (Tokyo, Japan: Nippon Electric Co., Ltd.).

72. R. L. Paul, "Manipulator Path Control," Proc. Int'l Conf. on Cyb. & Soc., San Francisco, CA, 23-25 Sep. 1975 (Piscataway, NJ: IEEE Svc. Ctr.), pp. 147-152.

73. C. A. Rosen, D. Nitzan, et al., "Machine Intelligence Research Applied to Industrial Automation," 6th Report, NSF Grant APR-75-13074, SRI Project 4391, (Menlo Park, CA: SRI Int'l, Artif. Intel. Ctr., Nov. 1976).

74. R. Finkel, "Constructing and Debugging Manipulator Programs," Memo AIM-284 and Report CS-567, (Stanford, CA: Stanford Artif. Intel. Proj., Aug. 1976), 171 pp.

75. W. B. Heginbotham, M. Dooner, and D. N. Kennedy, "Computer Graphics Simulation of Industrial Robot Interaction," Proc. 7th Int'l Symp. on Indust. Robots, Tokyo, 19-21 October, 1977 (Tokyo: Japan Indust. Robot Assoc.), pp. 169-176.

76. W. B. Heginbotham, M. Dooner, and D. Kennedy, "Analysis of Industrial Robot Behaviours by Computer Graphics," in Proc. of 3rd CISM-IFTOMM Symp. on Theory and Practice of Robots and Manipulators, Udine, Italy, 12-15 Sep. 1978, (New York: Springer-Verlag, 1978).

77. W. B. Heginbotham, M. Dooner, and D. N. Kennedy, "Rapid Assessment of Industrial Robots Performance by Interactive Computer Graphics," Proc. 9th Int'l Symp. on Indust. Robots, Wash., D.C., 13-15 March, 1979 (Dearborn, MI: Soc. Manuf. Eng.'s, Mkt. Svc. Dept.), pp. 563-574.

78. W. B. Heginbotham, M. Dooner, and K. Case, "Assessing Robot Performance with Interactive Computer Graphics," Robotics Today (Winter 1979-80), pp. 33-35.

79. J. Katowski, "Remote Manipulators as a Computer Input Device," (Chapel Hill, NC: U. N. C., Dept. Comp. Sci., 1971).

80. M. C. Bonney, et al., "Using SAMMIE for Computer Aided Workplace and Work Task Design," Proc. 6th Congress, Intl. Ergonomics Assoc. (MD: U. Maryland, 1976).

81. R. P. Burton, "Real-Time Measurement of Multiple three-Dimensional Positions," Ph. D. Thesis, Order No. 73-21,248 (Salt Lake City, UT: U. Utah, Comp. Sci. Dept., 1973), 129 pp.

82. G. Macri, "Anaylsis of First UTD Installation Failures," Paper MS77-735 (Dearborn, MI: Soc. Manuf. Eng.'s, Mkt. Svc. Dept., 1978).

83. N. Heroux and G. Munson, Jr., "Robots Reduce Exposure to Some Industrial Hazards," in Industrial Robots, Vol. I, Fundamentals, (Dearborn, MI: Soc. Manuf. Eng.'s, Mkt. Svc. Dept., 1979), pp. 109-115.

Appendix A
Glossary of Terms

Categories

1. General robotics terms
2. Related technical areas
3. Types of robots
4. Applications
5. Mechanical hardware
6. Performance measures
7. Statics and kinematics
8. Dynamics and control
9. Sensory feedback
10. Computer and control hardware
11. Software
12. Operator interfaces
13. Communications
14. Economic analysis

---

* National Engineering Laboratory, National Bureau of Standards, Washington D.C., 20234, April 1980.

1. GENERAL ROBOTICS TERMS

ADAPTABLE - See definition 2 of FLEXIBLE.

Capable of making self-directed corrections. In a robot, this is often accomplished with the aid of visual, force, or tactile sensors.

ARCHITECTURE - Physical and logical structure of a computer or manufacturing process.

AUTOMATION - The theory, art, or technique of making a process automatic, self-moving, or self-controlling.

CONTROL HIERARCHY - A relationship of control elements whereby the results of higher-level control elements are used to command lower-level elements.

FLEXIBLE - Pliable or capable of bending. In robot mechanisms this may be due to joints, links, or transmission elements. Flexibility allows the end point of the robot to sag or deflect under load and to vibrate as a result of acceleration or deceleration.

Multipurpose; adaptable; capable of being redirected, retrained or used for new purposes. Refers to the reprogrammability or multi-task capability of robots.

HIERARCHY - A relationship of elements in a structure divided into levels with those at higher levels having priority or precedence over those at lower levels (see control hierarchy and sensory hierarchy).

INTERFACES - A shared boundary. An interface might be a mechanical or electrical connection between two devices; it might be a portion of computer storage accessed by two or more programs; or it might be a device for communication to or from a human operator.

LEVEL OF AUTOMATION - The degree to which a process has been made automatic. Relevant to the level of automation are questions of automatic failure recovery, the variety of situations that will be automatically handled, and the situation under which manual intervention or action by humans is required.

MANIPULATION - The operation of grasping and moving an object.

MANIPULATOR - A mechanism, usually consisting of a series of segments, jointed or sliding relative to one another, for the purpose of grasping and moving objects usually in several degrees of freedom. It may be remotely controlled by a computer or by a human.

MODULAR - Made up of subunits which can be combined in various ways.

In robots, a robot constructed from a number of interchangeable subunits, each of which can be one of a range of sizes or have one of several possible motion styles (prismatic, cylindrical etc.) and number of axes.

"Modular design" permits assembly of products, or software or hardware from standardized components.

PROGRAMMABLE - Capable of being instructed to operate in a specified manner or of accepting set points or other commands from a remote source.

PROGRAMMABLE MANIPULATOR - A device that is capable of manipulating objects by executing a stored program resident in its memory.

REDUNDANCY - Duplication of information or devices in order to improve reliability.

ROBOT - A mechanical device that can be programmed to perform some task of manipulation or locomotion under automatic control.

SENSORY HIERARCHY - A relationship of sensory processing elements whereby the results of lower-level elements are utilized as inputs by higher-level elements.

2. RELATED TECHNICAL AREAS

ADAPTIVE CONTROL - A control method in which control parameters are continuously and automatically adjusted in response to measured process variables to achieve near-optimum performance.

ARTIFICIAL INTELLIGENCE - The ability of a device to perform functions that are normally associated with human intelligence, such as reasoning, planning, problem solving, pattern recognition, perception, cognition, understanding, and learning.

COMPUTER-AIDED DESIGN (CAD) - The use of a computer to assist in the creation of modification of a design.

COMPUTER-AIDED MANUFACTURE (CAM) - The use of computer technology in the management, control, and operation of manufacturing.

COMPUTER-MANAGED PARTS MANUFACTURE (CMPM) - Computer-aided manufacture of discrete parts, usually when a number of processing and product transport operations are coordinated by computer.

COMPUTER NUMERICAL CONTROL (CNC) - The use of a dedicated computer within a numerical control unit with the capability of local data input. It may become part of a DNC system by direct link to a central computer.

DIRECT DIGITAL CONTROL (DDC) - Use of a computer to provide the computations for the control functions of one or multiple control loops used in process control operations.

DIRECT NUMERICAL CONTROL (DNC) - The use of a computer for distribution of part program data via data lines to a plurality of remote NC machine tools.

DISTRIBUTED CONTROL - A control technique whereby portions of a single control process are located in two or more places.

FLEXIBLE MANUFACTURING SYSTEM - An arrangement of machines (usually NC machining centers with tool changers) interconnected by a transport system. The transporter carries work to the machines on pallets or other interface units so that accurate work-machine registration is rapid and automatic. A central computer controls machines and transport. May have a variety of parts being processed at any one time.

GROUP TECHNOLOGY - A system for coding parts based on similarities in geometrical shape or other characteristics of the parts.

The grouping of parts into families based on similarities in their production so that the parts of a particular family could then be processed together.

HIERARCHICAL CONTROL - A distributed control technique in which the controlling processes are arranged in a hierarchy. See HIERARCHY.

NUMERICAL CONTROL (NC) - A technique that provides prerecorded information in a symbolic form representing the complete instructions for the operation of a machine.

PART CLASSIFICATION - A coding scheme, typically involving four or more digits, that specifies a discrete product as belonging to a part family.

PATTERN RECOGNITION - Description or classification of pictures or other data structures into a set of classes or categories; a subset of the subject artificial intelligence.

TRANSFER MACHINE - An apparatus or device for grasping a workpiece and moving it automatically through stages of a manufacturing process.

3. TYPES OF ROBOTS

ANDROID - A robot that approximates a human in physical appearance.

ASSEMBLY ROBOT - A robot designed, programmed, or dedicated to putting together parts into subassemblies or complete products.

BANG-BANG ROBOT - A robot in which motions are controlled by driving each axis or degree of freedom against a mechanical limit stop. See also FIXED-STOP ROBOT.

BILATERAL MANIPULATOR - A master-slave manipulator with symmetric force reflection where both master and slave arms have sensors and actuators such that in any degree of freedom a positional error between the master and slave results in equal and opposing forces applied to the master and the slave arms.

A two-armed manipulator (can refer to two arms performing a task in cooperative movements, or can refer to two arms in the sense of a master- slave manipulator).

CYLINDRICAL COORDINATE ROBOT - A robot whose manipulator arm degrees of freedom are defined primarily by cylindrical coordinates.

FIXED-STOP ROBOT - A robot with stop point control but no trajectory control. That is., each of its axes has a fixed limit at each end of its stroke and cannot stop except at one or the other of these limits. Such a robot with N degrees of freedom can therefore stop at no more than two locations (where location includes position and orientation). Often very good repeatibility can be obtained.

INDUSTRIAL ROBOT - A robot used for handling, processing, assembling, or inspecting materials or parts in manufacture or construction; usually used in a factory.

INTELLIGENT ROBOT - A robot that can make sophisticated decisions and behaviorial choices through its sensing and recognizing capabilities.

LIMITED-DEGREE-OF-FREEDOM ROBOT - A robot able to position and orient its end effector in fewer than six degrees of freedom.

MASTER-SLAVE MANIPULATOR - A class of teleoperator having geometrically isomorphic master and slave arms. The master is held and positioned by a person; the slave duplicates the motions, sometimes with a change of scale in displacement or force.

MATERIALS-HANDLING ROBOT - A robot designed, programmed, or dedicated to grasping, transporting, and positioning materials in the process of manufacture.

MATERIALS-PROCESSING ROBOT - A robot designed, programmed, or dedicated to cutting, forming, heat treating, finishing, or otherwise processing materials as part of manufacture.

MOBILE ROBOT - A robot mounted on a movable platform.

NUCLEAR TELEOPERATOR - A device used for manipulation or inspection operations in a radioactive environment, sometimes incorporating some mobility capability by means of a wheeled or tracked vehicle, and controlled continuously by a remote human operator.

OPEN-LOOP ROBOT - A robot that incorporates no feedback, i.e., no means of comparing actual output to commanded input of position or rate.

PICK-AND-PLACE ROBOT - A simple robot, often with only two or three degrees of freedom, that transfers items from place to place by means of point-to-point moves. Little or no trajectory control is available. Often referred to as a bang-bang robot.

PROSTHETIC ROBOT - A robotic device that substitutes for lost manipulative or mobility functions of the human limbs.

RECORD-PLAYBACK ROBOT - A manipulator for which the critical points along desired trajectories are stored in sequence by recording the actual values of the joint position encoders of the robot as it is moved under operator control. To perform the task, these points are played back to the robot servo system.

SENSORY-CONTROLLED ROBOT - A robot whose control is a function of information sensed from its environment.

SEQUENCE ROBOT - A robot whose motion trajectory follows a preset sequence of positional changes.

SERVO-CONTROLLED ROBOT - A robot driven by servomechanisms, i.e., motors whose driving signal is a function of the difference between commanded position and/or rate and measured actual position and/or rate. Such a robot is capable of stopping at or moving through a practically unlimited number of points in executing a programmed trajectory.

SPACE ROBOT - A robot used for manipulation or inspection in an earth orbit or deep space environment.

SPHERICAL COORDINATE ROBOT - A robot whose manipulator arm degrees of freedom are defined primarily by spherical coordinates.

SUPERVISORY-CONTROLLED ROBOT - A robot incorporating a hierarchical control scheme, whereby a device having sensors, actuators, and a computer, and capable of autonomous decision making and control over short periods and restricted conditions, is remotely monitored and intermittently operated directly or reprogrammed by a person.

TELEOPERATOR - A device having sensors and actuators for mobility and/or manipulation, remotely controlled by a human operator. A teleoperator allows an operator to extend his sensory-motor function to remote or hazardous environment.

UNDERSEA TELEOPERATOR - A device used for manipulation or inspection operations undersea; usually part of a mobile submarine vehicle.

4. APPLICATIONS

BATCH MANUFACTURE - The production of parts in discrete runs or batches, interspersed with other production operations or runs of other parts.

CELL - An ICAM manufacturing unit consisting of a number of work stations and the materials transport mechanisms and storage buffers which interconnect them.

CELL CONTROL - A module in the ICAM control hierarchy that controls a cell. The cell control module is controlled by a center control module, if one exists. Otherwise it is controlled by the factory control level.

CENTER - An ICAM manufacturing unit consisting of a number of cells and the materials transport and storage buffers that interconnect them.

CENTER CONTROL - A module in the ICAM control hierarchy that controls a center. The center control module is controlled by the factory control level.

FACTORY - An ICAM manufacturing unit consisting of a number of centers and the materials transport, storage buffers, and communications that interconnect them.

FACTORY CONTROL - A module in the ICAM control hierarchy that controls a factory. Factories are controlled by management personnel and policies.

FIXTURE - A device to hold and locate a workpiece during inspection or production operations.

INTERLOCK - A device to prevent a machine from initiating further operations until some condition or set of conditions are fulfilled.

JIG - A device that holds and locates a workpiece but also guides, controls, or limits one or more cutting tools.

JOB SHOP - A discrete parts manufacturing facility characterized by a mix of products of relatively low-volume production in batch lots.

LOCATING SURFACES - Machined surfaces on a part which are used as reference surfaces for precise locating and clamping of the part in a fixture.

MACHINING CENTER - A numerically controlled machine tool, such as a milling machine, capable of performing a variety of operations such as milling, drilling, tapping, reaming, boring, etc. Usually also included are arrangements for storing 10 to 100 tools and mechanisms for automatic tool change.

MASS PRODUCTION - The large-scale production of parts in a continuous process uninterrupted by the production of other parts.

OPERATION - A single defined action.

PART FAMILY - A set of discrete products that can be produced by the same sequence of machining operations. This term is primarily associated with group technology.

PART ORIENTATION - The angular displacement of a product being manufactured relative to a coordinate system referenced to a production machine, e.g., a drilling or milling axis. Reorientation is often required as the product proceeds from one processing step to another.

PROCESS - A systematic sequence of operations to produce a specified result.

ROUTING - In production, the sequence of operations to be performed in order to produce a part or an assembly.

In telecommunications, the assignment of the communications path by which a message can reach its destination.

STATION CONTROL - A module in the ICAM control hierarchy that controls a work station. The station control module is controlled by a cell control module.

WORK-IN-PROCESS - Products in various stages of completion throughout the production cycle, including raw material that has been released for initial processing and finished products awaiting final inspection and acceptance for shipment to a customer.

WORK STATION - An ICAM manufacturing unit consisting of one or more numerically controlled machine tools serviced by a robot.

5. MECHANICAL HARDWARE

ACTUATOR - A motor.

A transducer that converts electrical, hydraulic, or pneumatic energy to effect motion of the robot.

ARM - An interconnected set of links and powered joints comprising a manipulator and supporting or moving a hand or end effector.

BASE - The platform or structure to which a robot arm is attached; the end of a kinematic chain of arm links and joints opposite to that which grasps or processes external objects.

BEVEL GEARS - Mating gears having conical external shapes whose axes of rotation are nonparallel.

CABLE DRIVE - Transmission of power from an actuator to a remote mechanism by means of flexible cable and pulleys.

CHAIN DRIVE - Transmission of power from an actuator to a remote mechanism by means of flexible chain and mating-toothed sprocket wheels.

EFFECTOR - An actuator, motor, or driven mechanical device.

END EFFECTOR - An actuator, gripper, or driven mechanical device attached to the end of a manipulator by which objects can be grasped or otherwise acted upon.

GRIPPER - A manipulator hand.

A device by which a robot may grasp and hold external objects

HAND - A device attached to the end of a manipulator arm, having a mechanism with closing jaws or other means to grasp objects.

HYDRAULIC MOTOR - An actuator consisting of interconnected valves or pistons that convert high-pressure hydraulic or pneumatic fluid into mechanical shaft rotation.

INDUCTION MOTOR - An alternating current motor in which torque is produced by the reaction between a varying or rotating magnetic field that is generated in stationary field magnets and the current that is induced in the coils or circuits of the rotor.

JOINT - Rotatory or linear articulation; axis of rotational or translational (sliding) degree-of-freedom of manipulator arm.

LEAD SCREW - A precision machine screw which, when turned, drives a sliding nut or mating part in translation.

LIMIT SWITCH - An electrical switch positioned to be actuated when a certain motion limit occurs, thereby to deactivating the actuator causing that motion.

PLANETARY DRIVE - A gear reduction arrangement consisting of a sum spur gear, two or more planetary spur gears, and an internally toothed ring gear.

POWER CYLINDER - A linear mechanical actuator consisting of a piston in a cylindrical volume and driven by high-pressure hydraulic or pneumatic fluid.

SERVOMECHANISM - An automatic control mechanism consisting of a motor driven by a signal that is a function of the difference between commanded position and/or rate and measured actual position and/or rate.

SERVOVALVE - A transducer whose input is a low-energy signal and whose output is a higher energy fluid flow that is proportional to the low-energy signal.

SHOULDER - The manipulator arm linkage joint that is attached to the base.

SOLENOID - A cylindrical coil of wire surrounding a movable core, which, when energized, sets up a magnetic field and draws in the core.

STEPPING MOTOR - An electric motor whose windings are arranged in such a way that the armature can be made to step in discrete rotational increments (typically 1/200th of a revolution) when a digital pulse is applied to an accompanying "driver" circuit. The armature displacement will stay locked in this angular position independent of applied torque, up to a limit.

STOP - A mechanical constraint or limit on some motion which can be set to stop the motion at a desired point.

TAPE DRIVE - Transmission of power from an actuator to a remote mechanism by means of flexible tape and pulleys.

TRANSDUCER - A device that converts one form of energy into another form of energy.

WORM GEAR - A short screw that mates to a gear whose axis of rotation is perpendicular to and offset from that of the worm screw. When the screw is turned, it drives the gear in rotation.

WRIST - The manipulator arm joint to which a hand or end effector is attached.

6. PERFORMANCE MEASURES

ACCURACY - Quality, state, or degree of conformance to a recognized standard.

Difference between the actual position response and the target position desired or commanded of an automatic control system.

CALIBRATION - The act of determining, marking, or rectifying the capacity or scale graduations of a measuring instrument or replicating machine.

To determine the deviation from standard so as to ascertain the proper correction factors.

DRIFT - The tendency of a system's response to move gradually away from the desired response.

DYNAMIC ACCURACY - Deviation from true value when relevant variables are changing with time.

Difference between actual position response and position desired or commanded of an automatic control system as measured during motion.

FAIL-SAFE - Failure of a device without danger to personnel or major damage to product or plant facilities.

FAIL SOFT - Same as GRACEFUL FAILURE.

GRACEFUL DEGRADATION - Decline in performance of some component part of a system without immediate and significant decline in performance of the system as a whole and/or decline in the quality of the product.

GRACEFUL FAILURE - Failure in performance of some component part of a system without immediate major interruption or failure of performance of the system as a whole and/or sacrifice in quality of the product.

LINEARITY - The degree to which an input/output relationship is propositional.

The degree to which a motion intended to be in a straight line conforms to a straight line.

LOAD - In physics, the external force applied to a body, or the energy required; also, the act of applying such force or requiring such energy.

In programming, to enter data into storage or working registers.

In production control, the amount of scheduled work planned for a manufacturing facility, often expressed as hours of work.

LOAD CAPACITY - The maximum weight or mass of a material that can be handled by a machine or process without failure.

LONG-TERM REPEATABILITY - Closeness of agreement of position movements, repeated under the same conditions during a long time interval, to the same location.

MAXIMUM SPEED - The greatest rate at which an operation can be accomplished according to some criterion of satisfaction.

The greatest velocity of movement of a tool or end effector that can be achieved in producing a satisfactory result.

MEAN-TIME-BETWEEN-FAILURE (MTBF) - The average time that a device will operate before failure.

MEAN-TIME-TO-REPAIR (MTTR) - The average time that a device is expected to be out of service after failure.

NET LOAD CAPACITY - The additional weight or masss of a material that can be handled by a machine or process without failure over and above that required for a container, pallet, or other device that necessarily accompanies the material.

PAYLOAD - The maximum weight or mass of a material that can be handled satisfactorily by a machine or process in normal and continuous operation.

PERFORMANCE - The quality of behavior.

The degree to which a specified result is achieved.

A quantative index of such behavior or achievement, such as speed, power, or accuracy.

PLAYBACK ACCURACY - Difference between a position command recorded in an automatic control system and that actually produced at a later time when the recorded position is used to execute control.

Difference between actual position response of an automatic control system during a programming or teaching run and that corresponding response in a subsequent run.

POSITION ERROR - In a servomechanism that operates a manipulator joint, the difference between the actual position of that joint and the commanded position.

PRECISION - The standard deviation or root-mean-squared deviation of values around their mean.

RATED LOAD CAPACITY - A specified weight or mass of a material that can be handled by a machine or process that allows for some margin of safety relative to the point of expected failure.

RELIABILITY - The probability that a device will function without failure over a specified time period or amount of usage.

REPEATABILITY - Closeness of agreement of repeated position movements, under the same conditions, to the same location.

RESOLUTION - The least interval between two adjacent discrete details that can be distinguished from one another.

The smallest increment of distance that can be read and acted upon by an automatic control system.

SHORT-TERM REPEATABILITY - Closeness of agreement of position movements, repeated under the same conditions during a short time interval, to the same location.

SPEED - The maximum speed at which the robot can move. Usually, the maximum tool tip speed in an inertial reference frame.

SPEED-PAYLOAD TRADEOFF - The relationship between corresponding values of maximum speed and payload with which an operation can be accomplished to some criterion of satisfaction, with all other factors remaining the same. See MAXIMUM SPEED and PAYLOAD.

SPEED-RELIABILITY TRADEOFF - The relationship between corresponding values of maximum speed and reliability with which an operation can be accomplished to some criterion of satisfaction, with all other factors remaining the same. See MAXIMUM SPEED and RELIABILITY.

SPRINGBACK - The deflection of a body when external load is removed. Usually refers to deflection of the end effector of a manipulator arm.

STATIC ACCURACY - Deviation from time value when relevant variables are not changing with time.

Difference between actual position response and position desired or commanded of an automatic control system as determined in the steady state, i.e., when all transient responses have died out.

STEADINESS - Relative absence of high-frequency vibration or jerk.

STRENGTH - Same as LOAD CAPACITY.

VELOCITY ERROR - In a servomechanism that operates a manipulator joint, the difference between the rate of change of the actual position of that joint and the rate of change of the commanded position.

7. STATICS AND KINEMATICS

AZIMUTH - Direction of a straight line to a point in a horizontal plane, expressed as the angular distance from a reference line, such as the observer's line of view.

BACKLASH - Free play in a power transmission system such as a gear train, resulting in a characteristic form of hysteresis.

CARTESIAN COORDINATE SYSTEM - A coordinate system whose axes or dimensions are three intersecting perpendicular straight lines and whose origin is the intersection.

CENTER OF ACCELERATION - That point in a rigid body around which the entire mass revolves.

CENTER OF GRAVITY - That point in a rigid body at which the entire mass of the body could be concentrated and produce the same gravity resultant as that for the body itself.

COMPLIANCE - The quality or state of bending or deforming to stresses within the elastic limit.

The amount of displacement per unit of applied force.

CYLINDRICAL COORDINATE SYSTEM - A coordinate system consisting of one angular dimension and two linear dimensions. These three coordinates specify a point on a cylinder.

DEAD BAND - A range within which a nonzero input causes no output.

DEGREE OF FREEDOM - One of a limited number of ways in which a point or a body may move or in which a dynamic system may change, each way being expressed by an indepdendent variable and all required to be specified if the physical state of the body or system is to be completely defined.

DISTAL - Away from the base, toward the end effector of the arm.

DROOP - Same as STATIC LOAD DEFLECTION.

ELEVATION - Direction of a straight line to a point in a vertical plane, expressed as the angular distance from a reference line, such as the observer's line of view.

END-POINT RIGIDITY - The resistance of the hand, tool, or end point of a manipulator arm to motion under applied force.

EXTENSION - Orientation or motion toward a position where the joint angle between two connected bodies is 180 degrees.

FIXED COORDINATE SYSTEM - A coordinate system fixed in time.

FLEXION - Orientation or motion toward a position where the joint angle between two connected bodies is small.

HYSTERESIS - The lagging of a physical response of a body behind its cause.

The asymmetry of the force/displacement relationship in one direction compared to that of another direction.

JOINT SPACE - The vector that specifies the angular or translational displacement of each joint of a multi-degree-of-freedom linkage relative to a reference displacement for each such joint.

LINEAR - Direction or motion as defined by one or more straight lines.

A relationship between quantities such that they add in a simple or arithmetic fashion.

LOAD DEFLECTION - The difference in position of some point on a body between a nonloaded and an externally loaded condition.

The difference in position of some point on a body between a nonloaded and an externally loaded condition.

The difference in position of a manipulator hand or tool, usually with the arm extended, between a nonloaded condition (other than gravity) and an externally loaded condition. Either or both static and dynamic (inertial) loads may be considered.

PAN - Orientation of a view, as with video camera, in azimuth.

Motion in the azimuth direction.

PITCH - An angular displacement up or down as viewed along the principal axes of a body having a top side, especially along its line of motion.

The axial displacement of successive threads of a screw.

POLAR COORDINATE SYSTEM - Same as SPHERICAL COORDINATE SYSTEM, usually as applied to points in a plane (only one angular dimension and one linear dimension used). Two coordinates specify a point on a circle.

PRONATION - Orientation or motion toward a position with the back, or protected side, facing up or exposed.

PROXIMAL - Close to the base, away from the end effector of the arm.

RECTANGULAR COORDINATE SYSTEM - Same as CARTESIAN COORDINATE SYSTEM, usually as applied to points in a plane (only two axes used).

RELATIVE COORDINATE SYSTEM - A coordinate system whose origin moves relative to world or fixed coordinates.

REMOTE CENTER COMPLIANCE (RCC) - A compliant device used to interface a robot or other mechanical workhead to its tool or working medium. The RCC allows a gripped part to rotate about its tip or to translate without rotating when pushed laterally at its tip. The RCC thus provides general lateral and rotational float and greatly eases robot or other mechanical assembly in the presence of errors in parts, jigs, pallets, and robots. It is especially useful in performing very close clearance or interference insertions.

ROLL - The angular displacement around the principal axis of a body, especially its line of motion.

ROTATION - Movement of a body around an axis, i.e., such that (at least) one point remains fixed.

SATURATION - A range within which the output is constant regardless of input.

SPHERICAL COORDINATE SYSTEM - A coordinate system, two of whose dimensions are angles, the third being a linear distance from the point of origin. These three coordinates specify a point on a sphere.

STATIC DEFLECTION - Load deflection considering only static loads, i.e., excluding inertial loads. Sometimes static deflection is meant to include the effects of gravity loads.

STIFFNESS - The amount of applied force per unit of displacement of a compliant body.

SUPINATION - Orientation or motion toward a position with the front, or unprotected side, facing up or exposed.

TILT - Orientation of a view, as with a video camera, in elevation.

Motion in the elevation direction.

TRANSLATION - Movement of a body such that all axes remain parallel to what they were, i.e., without rotation.

TWIST - Rotational displacement around a reference line; same as ROLL.

WINDUP - Colloquial term describing the twisting of a shaft under torsional load -- so called because the twist usually unwinds, sometimes causing vibration or other negative effects.

WORKING ENVELOPE - The set of points representing the maximum extent or reach of the robot hand or working tool in all directions.

WORKING RANGE - Same as WORKING ENVELOPE.

The range of any variable within which the system normally operates.

WORKING SPACE OR VOLUME - The physical space bounded by the working envelope in physical space.

WORLD COORDINATES - The coordinate system referenced to the earth or the shop floor.

WORK COORDINATES - The coordinate system referenced to the work piece, jig, or fixture.

YAW - An angular displacement left or right viewed from along the principal axis of a body having a top side, especially along its line of motion.

8. DYNAMICS AND CONTROL

ACTIVE ACCOMMODATION - Integration of sensors, control, and robot motion to achieve alternation of a robot's preprogrammed motions in response to felt forces. If a wrist force sensor and resolved motion rate control are employed, then the felt force vector can be used as stimulus to create quite general changes in the velocity vector of the end point. This technique can be used to stop a robot when forces reach set levels, or perform force feedback tasks like insertions, door opening and edge tracing.

ANALOG CONTROL - Control involving analog signal processing devices (electronic, hydraulic, pneumatic, etc.)

BANDWIDTH - The range of frequencies contained in a time function.

The range of frequencies to which a dynamic system will respond.

The range of frequencies which a communication channel will pass.

BANG-BANG CONTROL - Control achieved by a command to the actuator that at any time tells it to operate either in one direction or the other with maximum energy.

BANG-BANG-OFF CONTROL - Control achieved by a command to the actuator which at any time tells it to operate either in one direction or the other with maximum energy or to do nothing.

BREAKAWAY FORCE - Same as STATIC FRICTION, though this term implies more strongly than static friction that the resistive force is not constant as relative velocity increases.

CENTRALIZED CONTROL - Control decisions for two or more control tasks at different locations made at a centalized location.

CLOSED-LOOP CONTROL - Control achieved by a closed feedback loop, i.e., by measuring the degree to which actual system response conforms to desired system response, and utlizing the difference to drive the system into conformance.

COMPENSATION - Logical operations employed in a control scheme to counteract dynamic lags or otherwise to modify the transformation between measured signals and controller output to produce prompt stable response.

COMPUTED PATH CONTROL - A control scheme wherein the path of the manipulator end point is computed to achieve a desired result in conformance to a given criterion, such as an acceleration limit, a minimum time, etc.

COMPUTER CONTROL - Control involving one or more electronic digital computers.

CONTINUOUS PATH CONTROL - A control scheme whereby the inputs or commands specify every point along a desired path of motion.

CONTROL - The process of making a variable or system of variables conform to what is desired.

A device to achieve such conformance automatically.

A device by which a person may communicate his commands to a machine.

CONTROLLER - A device to achieve control.

COORDINATED AXIS CONTROL - Control wherein the axes of the robot arrive at their respective end points simultaneously, giving a smooth appearance to the motion.

- Control wherein the motions of the axes are such that the end point moves along a prespecified type of path (line, circle, etc.). Also called end point control.

DAMPING - The absorption of energy, as viscous damping of mechanical energy, resistive damping of electrical energy.

A property of a dynamic system which causes oscillations to die out and makes the response of the system approach a constant value.

DELAY - The time between input and output of a pulse or other signal which undergoes normal distortion.

DERIVATIVE CONTROL - Control scheme whereby the actuator drive signal is proportional to the time derivative of the difference between the input (desired output) and the measured actual output.

DIGITAL CONTROL - Control involving digital logic devices that may or may not be complete digital computers.

DYNAMIC RANGE - The range of any dynamic property of a system.

END-POINT CONTROL - Any control scheme in which only the motion of the manipulator end point may be commanded and the computer can command the actuators at the various degrees of freedom to achieve the desired result.

ERROR SIGNAL - The difference between desired response and actual response.

FEEDBACK - Use of the error signal to drive the control actuator.

FREQUENCY RESPONSE - The response of a dynamic system to a sinusoid.

The characterization of response of a dynamic system to any periodic signal according to the Fourier coefficients or the gain and phase at each frequency multiple of the period.

The characterization of dynamic response to a continuous spectral input according to a continuous plot of gain and phase as a function of frequency.

FRICTION - The rubbing of one body against another.

The resistive forces resulting from two bodies sliding relative to one another or from a body moving through a fluid.

INERTIA - The tendency of a mass at rest to remain at rest and of a mass in motion to remain in motion.

The Newtonian property of a physical mass that a force is required to change the velocity proportional to the mass and the time rate of change of velocity.

INTEGRAL CONTROL - Control scheme whereby the signal which drives the actuator equals the time integral of the difference between the input (desired output) and the measured actual output.

LAG - The tendency of the dynamic response of a passive physical system to respond later than desired.

The time parameter characterizing the transient response of a first order exponential system to a step.

The phase difference between input and response sinusoids.

Any time parameter which characterizes the delay of a response relative to an input.

The time it takes a signal or an object to move from one location to another (DELAY is a more precise term for this).

LEARNING CONTROL - A control scheme whereby experience is automatically used to provide for better future control decisions than those in the past.

MODERN CONTROL - A general term used to encompass both the description of systems in terms of state variables, canonical state equations, and the ideas of optimal control.

MULTIPROCESSOR CONTROL - Two or more control subtracks of the same overall control system accomplished simultaneously by more than one CPU.

NOISE - A spurious, unwanted, or distrubing signal.

A signal having energy over a wide range of frequencies.

OBJECTIVE FUNCTION - An equation defining a scalar quantity (to be minimized under given constraints by an optional controller) in terms of such performance variables as error, energy, and time. The objective function defines a trade-off relationship between these cost variables.

OPEN-LOOP CONTROL - Control achieved by driving control actutors with a sequence of preprogrammed signals without measuring actual system response and closing the feedback loop.

OPTIMAL CONTROL - A control scheme whereby the system response to a commanded input is optimal according to a specified objective function or criterion of performance, given the dynamics of the process to be controlled and the constraints on measuring.

OVERSHOOT - The degree to which a system response, such as change in reference input, goes beyond the desired value.

PASSIVE ACCOMMODATION - Compliant behavior of a robot's end point in response to forces exerted on it. No sensors, controls, or actuators are involved. The remote center compliance provides this in a coordinate system acting at the tip of a gripped part.

Use of the remote center compliance to achieve some of the capabilities of active accommodation.

POINT-TO-POINT CONTROL - A control scheme whereby the inputs or commands specify only a limited number of points along a desired path of motion. The control system determines the intervening path segments.

POSITION CONTROL - Control system in which the input (desired output) is the position of some body.

PROCESS CONTROL - Control of processes such as oil refining, chemical manufacture, water supply, and electrical power generation wherein the product and associated variables tend to be continuous in time.

PROPORTIONAL CONTROL - Control scheme whereby the signal that drives the actuator equals the difference between the input (desired output) and measured actual output.

PROPORTIONAL-INTEGRAL-DERIVATIVE CONTROL (PID) - Control scheme whereby the signal which drives the actuator equals a weighted sum of the difference, time integral of the difference, and time derivative of the difference between the input (desired output) and the measured actual output.

RATE CONTROL - Control system in which the input is the desired velocity of the controlled object.

RESOLVED MOTION RATE CONTROL - A control scheme whereby the velocity vector of the end point of a manipulaor arm is commanded and the computer determines the joint angular velocities to achieve the desired result.

Coordination of a robot's axes so that the velocity vector of the end point is under direct control. Motion in the coordinate system of the end point along specified directions or trajectories (line, circle, etc.) is possible. Used in manual control of manipulators and as a computational method for achieving programmed coordinate axis control in robots.

SENSORY CONTROL - Control of a robot based on sensor readings. Several types can be employed: Sensors used in threshold tests to terminate robot activity or branch to other activity; sensors used in a continuous way to guide or direct changes in robot motions (see ACCOMMODATION); sensors used to monitor robot progress and to check for task completion or unsafe conditions; and sensors used to retrospectively update robot motion plans prior to the next cycle.

SETTLING TIME - The time for a damped oscillatory response to decay to within some given limit.

SLEW RATE - The maximum velocity at which a manipulator joint can move; a rate imposed by saturation somewhere in the servo loop controlling that joint (e.g., by a valve's reaching its maximum open setting).

The maximum speed at which the tool tip can move in an inertial Cartesian frame .

STATIC FRICTION - The force required to commence the sliding of two bodies contacting relative to one another.

STEADY STATE - General term referring to a value that is not changing in time.

Response of a dynamic system due to its characteristic behavior, i.e., after any transient response has stopped; the steady-state response is either a constant or periodic signal.

STICTION - Same as STATIC FRICTION.

SUPERVISORY CONTROL - A control scheme whereby a person or computer monitors and intermittently reprograms, sets subgoals, or adjusts control parameters of a lower level automatic controller, while the lower level controller performs the control task continuously in real time.

TIME CONSTANT - Any of a number of parameters of a dynamic function that have units of time.

Parameters that particularly characterize the temporal properties of a dynamic function, such as the period of a periodic function or the inverse of the initial slope of a first order exponential response to a step.

TRACKING - Continuous position control response to a continuously changing input.

TRANSIENT - General term referring to a value that changes in time.

Response of a dynamic system to a transient input such as a step or a pulse.

UNDERSHOOT - The degree to which a system response to a step changes in reference input falls short of the desired value.

VISCOUS FRICTION - The resistive force on a body moving through a fluid.

Ideally, a resistive force proportional to relative velocities of two sliding bodies, or of a body and a fluid.

9. SENSORY FEEDBACK

ACTIVE ILLUMINATION - Illumination that can be varied automatically to extract more visual information from a scene, e.g., by turning lamps on and off, by adjusting brightness, by projecting a pattern on objects in the scene, or by changing the color of the illumination.

BINARY PICTURE - A digitized image in which the brightness of the pixels can have only two different values, such as white or black, zero, or one.

CCD CAMERA - A solid-state camera that uses a CCD (Charge-Coupled Device; also called a bucket brigade device) to transform a light image into a digitized image.

A CCD is similar to a CID, except that its method of operation forces readout of pixel brightnesses in a regular line-by-line scan pattern. There is only one readout station, and charges are shifted along until they reach it.

CID CAMERA - A solid-state camera that uses a CID (Charge Injection Imaging Device) to transform a light image into a digitized image.

The light image focused on the CID generates minority carriers in a silicon wafer, which are then trapped in potential wells under metallic electrodes held at an elevated voltage. Each electrode corresponds to one pixel of the image.

To register the brightness of one pixel of the image, the voltage on the electrode that corresponds to that pixel is changed to inject the charge stored under that electrode into the substrate. This produces a current flow in the substrate that is proportional to the brightness of the image at that pixel location, and is therefore capable of producing a grey-scale image.

In a CID camera, pixels of the image can be read out in an arbitrary sequence. This is not possible with a CCD camera. In some CID cameras, the same image can be read out hundreds or thousands of times (nondestrutive readout capability).

CONDUCTIVE RUBBER - A material consisting of carbon granules suspended in rubber, whose electrical resistance decreases gradually as it is mechanically compressed.

CONTACT SENSOR - A device capable of sensing mechanical contact of the hand or some other part of the robot with an external object.

ENCODER - A type of transducer commonly used to convert angular or linear position to digital data.

EXTERNAL SENSOR - A sensor for measuring displacements, forces, or other variables in the environment external to the robot.

EXTEROCEPTOR - External sensor, usually used in physiology.

FORCE SENSOR - A sensor capable of measuring the forces and torques exerted by a robot at its wrist. Such sensors usually contain six or more independent sets of strain gages plus amplifiers. Computer processing (analog or digital) converts the strain readings into three orthogonal torque readings in an arbitrary coordinate system. When mounted in the work surface, rather than the robot's wrist, such a sensor is often called a pedestal sensor.

FRAME BUFFER - An electronic device capable of storing a digitized image in a digital memory for later readout and processing.

GREY-SCALE PICTURE - A digitized image in which the brightness of the pixels can have more than two values, typically 128 or 256; requires more storage space and much more sophisticated image processing than a binary image, but offers potential for improved visual sensing.

INDUCTOSYN - Trademark for Farrand Controls resolver, in which an output signal is produced by inductive coupling between metallic patterns in two glass members separated by a small air space. Produced in both rotary and linear configurations.

INTERNAL SENSOR - A sensor for measuring displacements, forces, or other variables internal to the robot.

INTEROCEPTOR - Internal sensor, usually used in physiology.

LINEAR-ARRAY CAMERA - A tv camera (usually solid-state) with an aspect ratio of 1:n; today, n is typically 128, 256, or 512.

MATRIX-ARRAY CAMERA - A tv camera (usually solid-state) with an aspect ratio of n:m, where neither n nor m is 1; typically 128 by 128 today.

PHOTORESISTOR - A device for measuring light whose resistance changes as a function of incident light.

PIEZO ELECTRIC - The property of certain crystalline salts to change their electrical impedance as a function of mechanical pressure.

PIXEL - A picture element. A small region of a scene within which variations of brightness are ignored. The pixel is assigned a brightness level that is the average of the actual image brightnesses within it. Pixels are usually arranged in a rectangular pattern across the scene, although some research has been done with hexagonal grids.

POTENTIOMETER - An encoder based upon tapping the voltage at various points along a continuous electrical resistive element.

PROXIMITY SENSOR - A device that senses that an object is only a short distance (e.g., a few inches or feet) away, and/or measures how far away it is . Proximity sensors work on the principles of triangulation of reflected light, lapsed time for reflected sound, and others.

RESOLVER - A rotary or linear feedback device that converts mechanical motion to analog electric signals that represent motion or position.

RUN-LENGTH ENCODING - A data-compression technique for reducing the amount of information in a digitized binary image. It removes the redundancy that arises from the fact that such images contain large regions of adjacent pixels that are either all white or all black (i.e., black-white transitions are relatively infrequent). The brightness information is replaced by a sequence of small integers that tell how many consecutive black and white pixels are encountered while traversing each scan line. For grey-scale imagery, some compression can be achieved by considering the first n high-order bits of the brightness information to represent n different binary images and then transforming each into run-length format (the low-order bits will vary so much that there will be little redundancy to remove).

SEGMENTATION - Partitioning of a scene into subregions; in "windowing," for example, the portion of the scene outside a rectangular subregion is ignored to speed up image processing.

SENSOR - A transducer or other device whose input is a physical phenomenon and whose output is a quantitative measure of that physical phenomenon.

SHAFT ENCODER - An encoder used to measure shaft position.

SMART SENSOR - A sensing device whose output signal is contingent upon mathematical or logical operations and inputs other than from the sensor itself.

SOLID-STATE CAMERA - A tv camera that uses some sort of solid-state integrated circuit instead of a vacuum tube to change a light image into a video signal. Solid-state cameras have the following advantages over vacuum-tube cameras:

- Ruggedness
- Small size
- No high voltages
- Insensitive to image burn and lag; antibloom capability is possible with the proper readout technique.
- Potentially very low cost, characteristic of solid-state technology
- A spatially stable, precise geometry which effectively superimposes a fixed, repeatable measurement grid over the object under observation without the pin-cushion or barrel distortion introduced by the deflection systems of tube cameras.

STRAIN GAGE - A sensor that, when cemented to elastic materials, measures very small amounts of stretch by the change in its electrical resistance. When used on materials with high modulus of elasticity, strain gages become force sensors.

STRAIN-GAGE ROSETTE - Multiple strain gages cemented in two- or three-dimensional geometric patterns such that independent measurements of the strain on each can be combined to yield a vector measurement of strain or force.

STRUCTURED LIGHT - Illumination designed so that the three-dimensional pattern of light energy in the viewing volume causes visible patterns to appear on the surface of objects being viewed, from which patterns that are the shape of the objects can easily be determined.

SYNCHRO - A shaft encoder based upon differential inductive coupling between an energized rotor coil and field coils positioned at different shaft angles.

TACHOMETER - A rotational velocity sensor.

TACTILE SENSOR - A sensor that makes physical contact with an object in order to sense it; includes touch sensors, tactile arrays, force sensors, and torque sensors. Tactile sensors are usually constructed from microswitches, strain gages, or pressure-sensitive conductive elastomers.

TEMPLATE MATCHING - Pixel-by-pixel comparison of an image of a sample object with the image of a reference object; usually for purposes of identification, but also applicable to inspection.

THRESHOLDING - The process of quantizing pixel brightness to a small number of different levels (usually two levels, resulting in a binary image). A threshold is a level of brightness at which the quantized image brightness changes.

VIDECON - Trade name for a particular type of small vacuum tube used to change light images into video signals; a tv camera that contains such a tube.

10. COMPUTER AND CONTROL HARDWARE

ANALOG-TO-DIGITAL CONVERTER (A/D) - A hardware device that senses an analog signal and converts it to a representation in digital form.

CENTRAL PROCESSING UNIT (CPU) - The part of a computer that executes instructions and operates on data.

COMPLEMENTARY METAL-OXIDE SEMICONDUCTOR (CMOS) - An integrated circuit logic family characterized by very low power dissipation, moderate circuit density per chip, and moderate speed of operation.

CONTROLLER _ An information processing device whose inputs are both desired and measured position velocity or other pertinent variables in a process and whose outputs are drive signals to a controlling motor or actuator.

A communication device through which a person introduces commands to a control system.

A person who does the same.

DIGITAL-TO-ANALOG CONVERTOR (D/A) - A device that transforms digital data into analog data.

HOST COMPUTER - The primary or controlling computer in a multiple computer operation.

INPUT-OUTPUT (I/O) - Pertaining to either input or output signals or both.

A general term for the equipment used to communicate with a computer.

The data involved in such communication.

The media carrying the data for input-output.

INTEGRATED CIRCUIT (IC) - An electronic circuit packaged in a small unit ranging from 0.3 to 2 inches square, varying in complexity and function from simple logic gates to microprocessors, amplifiers, and analog-digital converters. The circuit may be constructed on a single semiconductor substrate, a configuration called monolithic, or several such circuits can be connected in one package called a hybrid.

LARGE SCALE INTEGRATION (LSI) - A classification for a scale of complexity of an integrated electronic circuit chip. Other classes are medium-scale integration (MSI) and small-scale integration (SSI).

MAGNETIC CORE MEMORY - A configuration of magnetic beads, strung on current-carrying conductors so as to retain a magnetic polarization for the purpose of storing and retrieving data.

MEMORY - A device into which data can be entered, in which it can be held, and from which it can be retrieved at a later time.

METAL-OXIDE SEMICONDUCTOR (MOS) - A semiconductor used by manufacturing technology to produce integrated circuit logic components.

MICROCOMPUTER - A computer that uses a microprocessor as its basic element.

MICROPROCESSOR - A basic element of a central processing unit constructed as a single integrated circuit. A microprocessor typically has a limited instruction set that may be expanded by microprogramming. A microprocessor may require additional circuits to become a central processing unit.

MULTIPLEXER - A hardware device that allows communication of multiple signals over a single channel by repetitively sampling each signal.

MULTIPROCESSOR - A computer that can execute one or more computer programs employing two or more processing units under integrated control of programs or devices.

OPERATIONAL AMPLIFIER - A high-gain amplifier used as the basic element in analog computation.

PERIPHERAL EQUIPMENT - Any unit of equipment, distinct from the central processing unit, which may provide the system with outside communication.

PROGRAMMABLE CONTROLLER - A controller whose algorithm for computing control outputs is programmable.

PROGRAMMABLE READ-ONLY MEMORY (PROM) - A read-only memory that can be modified by special electronic procedures.

RANDOM-ACCESS MEMORY (RAM) - A data storage device wherein the time required for obtaining data from or placing data into storage is independent of the location of the data most recently obtained or placed into storage.

READ-ONLY MEMORY (ROM) - A data-storage device generally used for control programs whose content is not alterable by normal operating procedures.

SILICON-CONTROLLED RECTIFIER (SCR) - An electronic device that is generally used in control systems for high-power loads such as electronic heating elements.

TRANSISTOR-TRANSISTOR LOGIC (TTL) - A common electronic logic configuration used in integrated circuits characterized by high speed and noise immunity.

11. SOFTWARE

ACCESS TIME - The time interval between the instant at which data are called for from a storage device and the instant delivery is completed.

ASSEMBLER - A program that translates symbolic codes into machine language and assigns memory locations for variables and constants.

ASSEMBLY LANGUAGE - An operation language, composed of brief expressions, that is translated by an assembler into a machine language. The language result (object code) from the assembler is a character-for-character translated version of the original.

BACKGROUND PROCESSING - The automatic execution of lower priority programs when higher priority programs are not using the system resources. Contrast with foreground processing.

BRANCHING - Transfer of control during program execution to an instruction other than the next sequential instruction. If the next instruction selected is predetermined, the branch is an unconditional branch; if the next instruction is selected on the basis of some sort of test, it is a conditional branch. A robot must possess the ability to execute conditional branches in order to react intelligently to its environment. The wider the variety of tests it can perform, the better it can react.

COMPILER - A program that converts a program written in a high-level language such as FORTRAN into binary coded instructions that the machine can interpret.

CONDITIONAL STATEMENT - A computer program step that specifies a dependence on whether certain tests of criteria are met.

CROSS-ASSEMBLER - A computer program to translate instructions into a form suitable for running on another computer.

DATA BASE - A collection of data fundamental to an enterprise; the data is comprised of comprehensive files of information having predetermined structure and organization and suitable for communication, interpretation or processing by humans or automatic means.

DEFAULT VALUE - A value that is used until a more valid one is found.

DIAGNOSTIC - A test or series of tests used to verify a system.

DOUBLE PRECISION - Pertaining to the use of two computer words to represent a number.

EDITOR - A routine that performs editing operations.

EXECUTE - To carry out an instruction or perform a routine.

FILE - A repository of organized information consisting of records, items or arrays, and data elements.

FIRMWARE - Programs loaded in read-only memory (ROM). Firmware is often a fundamental part of the system's hardware design, as contrasted to software, which is not fundamental to the hardware operation.

FIXED-POINT REPRESENTATION - A number system in which the position of the decimal point is fixed with respect to one end of the string of numerals, according to some convention.

FLOATING-POINT REPRESENTATION - A number representation system in which each number, as represented by a pair of numerals, equals one of those numerals times a power of an implicit fixed position integer base, where the power is equal to the implicit base raised to the exponent represented by the other numeral.

FOREGROUND PROCESSING - The automatic execution of programs that have been designed to preempt the use of computing facilities. Usually a real-time program. Contrast with BACKGROUND PROCESSING.

HEXADECIMAL - Pertaining to number system with a base of 16 (0-15).

HIGH-LEVEL LANGUAGE - Programming language that generates machine codes from problem or function-oriented statements. ALGOL, FORTRAN, PASCAL, and BASIC are four commonly used high-level languages. A single functional statement may translate into a series of instructions or subroutines in machine language, in contrast to a low-level (assembly) language in which statements translate on a one-for-one basis.

INSTRUCTION SET - The list of machine language instructions which a computer can perform.

INTERLOCK - To prevent a machine or device from initiating further operations until the operation in process is completed.

INTERPRETER - A program that translates and executes each source language expression before translating and executing the next one.

A routine which decodes instructions and produces a machine language routine to be executed at a later time.

INTERRUPT - To stop a process in such a way that it can be resume.

To get a computer system's attention especially for the purpose of input/output of data, for making an inquiry or receiving a reply, or for carrying out interactive processes or procedures.

LOOP - A sequence of instructions that is executed repeatedly until some specified condition is met.

MACHINE LANGUAGE - A language that is used directly by a machine.

MACRO - Programming with instructions (equivalent to a specified sequence of machine instructions) in a source language.

MEMORY PROTECTION - In data processing, an arrangement for preventing access to storage for either reading or writing or both.

MENU - A display of options on a terminal device for user selection.

MONITOR - Software or hardware that observes, supervises, controls, or verifies the operations of a system.

OFF-LINE - Pertaining to devices not under direct control of the central processing unit.

Operation where the CPU operates independently of the time base of input data or peripheral equipment.

ON-LINE - Pertaining to devices under direct control of the central processing unit.

Operation where input data is fed directly from the measuring devices into the CPU, or where data from the CPU is transmitted directly to where it is used. Such operation is in real time.

OPERATING SYSTEM - Software which controls the execution of computer programs and which may provide scheduling, debugging, input-output control, accounting, computation, storage assignment, data management, and related services.

PARALLEL PROCESSING - Concurrent or simultaneous execution of two or more operations, such as multiple arithmetic or logic units in devices..

PARITY CHECK - A check that tests whether the number of ones or zeros in an array of binary digits is even or odd. Such parity checks are widely used for paper tapes, magnetic tapes, and other computer memories.

REAL TIME - Pertaining to computation performed while the related physical process is taking place so that results of the computation can be used in guiding the physical process.

SOURCE PROGRAM - In a language, a program that is an input to a given translation process.

12. OPERATOR INTERFACES

ANALOGIC CONTROL - Pertaining to control by communication signals which are physically or geometrically isomorphic to the variables being controlled, usually by a human operator.

A device for effecting such control. Compare to SYMBOLIC CONTROL.

ANNUNCIATOR - A light or sound signal designed to attract attention.

CATHODE-RAY TUBE (CRT) - A device that presents data in visual form by means of a controlled electron beam impinging on a phosphorescent surface.

CURSOR - A moveable pointer, indicator, or marker on a visual display.

EXOSKELETON - An articulated mechanism whose joints correspond to those of a human arm, and, when attached to the arm of a human operator, will move in correspondence to his. Exoskeleton devices are sometimes instrumented and used for master-slave control of manipulators.

JOYSTICK - A moveable handle which a human operator may grasp and rotate to a limited extent in one or more degrees of freedom and whose variable position or applied force is measured, resulting in commands to a control system.

LIGHT-EMITTING DIODE (LED) - A semiconductor device that gives off light when current passes through it.

LIQUID CRYSTAL DISPLAY (LCD) - A display device consisting of liquid crystal material hermetically sealed between two glass plates that changes its optical properties under the influence of electrical current. One type of LCD depends upon ambient light for its operation, while a second depends upon a backlighting source.

REPLICA MASTER - A teleoperator control mechanism that is kinematically equivalent to the slave manipulator or other device that is being controlled (i.e., the master has the same kind of joints in the same relative positions as does the slave). As a human operator moves the master by hand, the control system forces the slave to follow the master's motions. A replica master may be larger than, smaller than, or the same size as the slave device it controls. The control system may reflect back to the joints of the master any forces and torques that are applied to corresponding joints of the slave (bilateral force feedback) to allow the operator to "feel" objects remotely through the slave. The replica master (and the slave) may or may not have a geometry similar to that of a human arm.

SCROLL - A graphic display technique whereby the generation of a new line of alphanumeric text at the bottom of a display screen automatically regenerates all other lines of text one line higher than before and deletes the top line.

STORAGE TUBE - A CRT display device which retains a graphic image if no new signals are imposed.

SWITCH CONTROL - Control of a machine by a person through movement of a switch to one of two or a small number of positions.

The device used for such control.

SYMBOLIC CONTROL - Pertaining to control by communication of discrete alpha-numeric or pictorial symbols that are not physically isomorphic with the variables being controlled, usually by a human operator.

A device for effecting such control. Compare to ANALOGIC CONTROL.

TEACH - To guide a manipulator arm through a series of points or in a motion pattern as a basis for subsequent automatic action by the manipulator.

TEACHING INTERFACE - The physical configuration of the machine or the devices by which a human operator teaches a machine. See TEACH.

THERMOCHROMIC DISPLAY - A display device consisting of materials that change to different colors when heated to different temperatures.

13. COMMUNICATIONS

ADCCP - Advanced Data Communication Control Procedure; an ANSI standard protocol for communication that is becoming increasingly popular in the United States; closely compatible with the HDLC protocol.

ASCII - American Standard Code for Information Interchange, a common coding scheme for alphanumeric characters and terminal control interfacing.

ACOUSTIC COUPLER - An electronic device that sends and receives digital data through a standard telephone handset. To transmit data, the digital signals are converted to audible tones that are acoustically coupled to a telephone handset. To receive data, the acoustically coupled audible signals are converted to digital signals.

ANALOG COMMUNICATIONS - Transfer of information by means of a continuously-variable quantity, such as the voltage produced by a strain gage or air pressure in a pneumatic line.

BAUD - A unit of signalling speed equal to the number of discrete conditions (bits) of signal events per second.

BISYNC - Binary Synchronous Communication Protocol; an early standard protocol for half-duplex communication, developed by IBM about 1965; in wide use today.

BUS - One or more conductors used for transmitting signals or power.

An information coding scheme by which different signals can be coded and identified when sharing a common data channel.

COMMUNICATIONS LINK - Any mechanism for the transmission of information; usually electrical. May be serial or parallel; synchronous or asynchronous; half duplex or full duplex; encrypted or clear; point-to-point, multidrop, or broadcast; may transmit binary data or text; may use standard characer codes to represent text and control information, such as the ASCII, EBCDIC, or BAUDOT (tty) codes; may use a handshaking protocol to synchronize operations of computers or devices at opposite ends of the link such as BISYNC, HDLC, or ADCCP.

DIGITAL COMMUNICATIONS - Transfer of information by means of a sequence of signals called bits (for BInary digiTS), each of which can have one of two different values. The signals may, for example, take the form of two different voltage levels on a wire or the presence of absence of light in a fiber optic light guide. Can be made arbitrarily insensitive to external disturbances by means of error control procedures.

ECHO CHECK - A method of checking the accuracy of transmission of data in which the received data are returned to the sending end for comparison with the original data.

ERROR CONTROL PROCEDURE - The inclusion of redundant information in a message (e.g., parity bits, check sums, cyclic redundancy check characters, use of Hamming codes, fire codes, etc.) to permit the detection (and in some cases the correction) of errors that arise from nosie or other disturbances in the transmission medium. May involve retransmission of messages until they are correctly received.

FULL DUPLEX - In communications, pertaining to simultaneous two-way independent transmission in both directions.

HDLC - High-level Data Link Control protocol. It is bit oriented, code independent, and suited to full-duplex communication. It has a potential of twice the throughput rate of BISYNC because it does not require immediate acknowledgments to each message frame. International Standard ISO 3309-1976 (E) defines in detail the frame structure to be used for each HDLC transmission as:

1. An 8-bit flag sequence (01111110)

2. An 8-bit secondary station address field

3. An 8-bit control field containing:

    a. commands from the primary station to the secondary
    b. responses from the secondary to the primary
    c. message sequence numbers

4. An optional information field of variable length

5. A 16-bit frame checking sequence

6. An 8-bit flag sequence (01111110)

HALF DUPLEX - In communications, pertaining to alternate, one-way-at-a-time transmissions.

MODULATOR-DEMODULATOR (MODEM) - An electronic device that sends and receives digital data using telecommunication lines. To transmit data, the digital signals are used to vary (modulate) an electronic signal that is coupled into the telecommunication lines. To receive data, the electronic signals are converted (demodulated) to digital data.

PARALLEL COMMUNICATIONS - A digital communication method that transmits the bits of a message several at a time (usually 8 or 16 bits at a time); usually only used over distances of a few feet with electrical cables as the transmission medium.

POLLING - A technique by which each of the terminals sharing a communications line is periodically interrogated to determine whether it requires servicing. The multiplexer or control station sends a poll which, in effect, asks the terminal selected, "Do you have anything to transmit?"

PROTOCOL - The rules for controlling data communications between devices in computer systems.

RS-232-C, RS-422, RS-423, RS-449 - Standard electrical interfaces for connecting peripheral devices to computers. EIA Standard RS-449, together with EIA Standards RS-422 and RS-423, are intended to gradually replace the widely-used EIA Standard RS-232-C as the specification for the interface between data terminal equipment (DTE) and data circuit-terminating equipment (DCE) employing serial binary data interchange. Designed to be compatible with equipment using RS-232-C, RS-449 takes advantage of recent advances in integrated circuit design, reduces crosstalk between interchange circuits, permits greater distances between equipments, and permits higher data signaling rates (up to 2 million bits per second). RS-449 specifies functional and mechanical aspects of the interface, such as the use of two connectors having 37 pins and 9 pins instead of a single 25-pin connector. RS-422 specifies the electrical aspects for wideband communication over balanced lines at data rates up to 10 million bits per second. RS-423 does the same for unbalanced lines at data rates up to 100,000 bits per second.

SERIAL COMMUNICATIONS - A digital communication method that transmits the bits of a message one at a time; the most common long-distance transmission method; suitable for use with cable, radio, or modulated light as the transmission medium.

14. ECONOMIC ANALYSIS

CYCLE TIME - The period of time from starting one machine operation to starting another (in a pattern of continuous repetition).

DUTY CYCLE - The fraction of time during which a device or system will be active or at full power.

FLOOR-TO-FLOOR TIME - The total time elapsed for picking up a part, loading it into a machine, carrying out operations, and unloading it (back to the floor, bin, pallet, etc.); generally applies to batch production.

Appendix B
List of Current Literature

Topics

Advanced Automation Research

Advanced Vision Research

Application Criteria and New Robotic Applications

Artificial Intelligence Research on Robots

Attitude of Unions Towards Robotization

Compliance and Accommodation Technology

Computer Graphics for Simulation of Robotic Operations

Current Practice and Commercial Systems for Industrial Vision

End Effectors, Robot Accessories and Actuator Technology

Industrial Automation Surveys and Research Summaries

Industrial Vision Research

Manipulator Control Systems and Techniques

Manipulator Design

Modern Robotic Practice

Programming Languages and Software

Safety

Sensor Technology and Applications

Standardization Issues in Robotics

Surveys of Artificial Intelligence

Teleoperator Systems and Techniques

Miscellaneous References

Advanced Automation Research

Barbera, A., et al., "Control Strategies for Industrial Robot Systems," Final Report, Publ. PB-283539/5GA (Wash., D.C.: Natl. Bur. Stds. 1976), 10 pp.

Barbera, A., "An Architecture for a Robot Hierarchical Control System," Special Publ. 500-23 (Wash., D.C.: Natl. Bur. Stds., Dec. 1977).

Barbera, A., et al., "Hierarchical Control of Robots Using Microcomputers," Proc. 9th Int'l Symp. on Indust. Robots, Wash., D.C., 13-15 March, 1979 (Dearborn, MI: Soc. Manuf. Eng.'s, Mkt. Svc. Dept.), pp. 405-422.

Bejczy, A. K., "Robot Arm Dynamics and Control," Tech. Memo. 33-669 (Pasadena, CA: Jet Propulsion Lab, Cal. Inst. Tech., Feb. 1974).

Binford, T. O., et al., "Exploratory Study of Computer Integrated Assembly Systems, Progress Report 1," limited distribution / not available, (Stanford, CA: Stanford Artif. Intel. Proj., Dec. 1974), 365 pp.

Binford, T. O., et al., "Exploratory Study of Computer Integrated Assembly Systems, Progress Report 2," limited distribution / not available, (Stanford, CA: Stanford Artif. Intel. Proj., Aug. 1976), 310 pp.

Binford, T. O., et al., "Exploratory Study of Computer Integrated Assembly Systems, Progress Report 3," Memo AIM-285.3 and Report CS-568,PPB-259 130/3WC, (Stanford, CA: Stanford Artif. Intel. Proj., Aug. 1976), 336 pp.

Binford, T. O., et al., "Exploratory Study of Computer Integrated Assembly Systems, Progress Report 4," Memo AIM-285.4 and Report STAN-CS-76-568 (Stanford, CA: Stanford Artif. Intel. Proj., June 1977), 250 pp.

Birk, J., Kelley, R., et al., "Robot Computations for Orienting Workpieces," 1st Report, Nat'l Sci. Found. Grant APR-74-13935 (Kingston, RI: U. Rhode Island, Jan 1976).

Birk, J., et al., "General Methods to Enable Robots with Vision to Acquire, Orient, and Transport Workpieces," 3rd Report, Nat'l Sci. Found. Grant APR-74-13935, (Kingston, RI: U. Rhode Island, Aug 1977).

Birk, J., et al., "General Methods to Enable Robots with Vision to Acquire, Orient, and Transport Workpieces," 4th Report, Nat'l Sci. Found. Grant APR-74-13935, (Kingston, RI: U. Rhode Island, July 1978).

Bolles, R. C., "Part Acquisition Using the SRI Vision Module," Proc. Comp. Software & Applications Conf., Chicago IL., 6-8 Nov., 1979, pp. 872-877.

Finkel, R., et al., "AL, A Programming System for Automation," Memo AIM-243 and Report CS-456, supported by Nat'l Sci. Found. Contract GI-42906 and Adv'd Res. Proj's Agency Contract DAHC-15-73-C-0435, (Stanford, CA: Stanford Artif. Intel. Proj., Nov. 1974).

Finkel, R., "Constructing and Debugging Manipulator Programs," Memo AIM-284 and Report CS-567, (Stanford, CA: Stanford Artif. Intel. Proj., Aug. 1976), 171 pp.

Grossman, D. D. and Taylor, R. H., "Interactive Generation of Object Models with a Manipulator," Nat'l Sci. Found. Grant G142906, Adv. Res. Proj. Agency Contract DAHC-15-73-C-0435, Memo AIM-274 or Report STAN-CS-75-536 (Stanford, CA: Stanford Artif. Intel. Proj., Dec. 1975).

Grossman, D. D., "Monte Carlo Simulation of Tolerancing in Discrete Parts Manufacturing and Assembly," Nat'l Sci. Found. Grant APR-74-01390-A02, Memo AIM-280 or Report STAN-CS-76-555 (Stanford, CA: Stanford Artif. Intel. Proj., May 1976).

Ishida, T., "Force Control in Co-ordination of Two Arms," 5th Int'l Jt. Conf. on Artificial Intelligence, Cambridge, MA, 22-25 Aug., 1977 (Pittsburgh, PA: Carnegie-Mellon U., Dept. of Comp. Sci.), pp. 717-722.

Kahn, E. M., "The Near-Minimum-Time Control of Open-Loop Articulated Kinematic Chains," Memo AIM-106, supported by Adv'd Res. Proj's Agency Contract SD-183, (Wash., D.C.: Natl. Tech. Info. Service, U. S. Dept. of Commerce), also available from (Stanford, CA: Stanford Artif. Intel. Proj.), 171 pp.

Kashioka, S., et al., "An Approach to the Integrated Intelligent Robot with Multiple Sensory Feedback: Visual Recognition Techniques," Proc. 7th Int'l Symp. on Indust. Robots, Tokyo, 19-21 October, 1977 (Tokyo: Japan Indust. Robot Assoc.), pp. 531-538.

Lewis, R. L., "Autonomous Manipulation on a Robot: Summary of Manipulator Software Functions," Memo 33-679 (Pasadena, CA: Jet Propulsion Lab, Cal. Inst. Tech., March 1974).

Liebermann, L. I. and Wesley, M. A., "AUTOPASS: A Very High Level Programming Language for Mechanical Assembler Systems," Report RC 5599 (#24205) (Yorktown Heights, NY: IBM T. J. Watson Res. Ctr., Aug. 1975).

Liebermann, L. I. and Wesley, M. A., "AUTOPASS: An Automatic Programming System for Computer-Controlled Mechanical Assembly," IBM J. Res. & Devel., Vol. 21 (July 1977), pp. 321-333.

Lozano-Perez, T., and Wesley, M., "An Algorithm for Planning Collision-Free Paths Amongst Polyhedral Obstacles," Report 7171 (Yorktown Heights, NY: IBM T. J. Watson Res. Ctr., June 1978), revised version in Comm. of the ACM, Vol. 22, No. 10 (Oct. 1979), pp. 560-570.

McGhie, D., and Hill, J., "Vision-Controlled Subassembly Station," Paper MS78-685 (Dearborn, MI: Soc. Manuf. Eng.'s, Mkt. Svc. Dept., Nov. 1978), and in Proc. Robots III Conf., Chicago, IL, 7-9 Nov. 1978, supported by Nat'l Sci. Found. Grant APR75-13074, reprints available (Menlo Park, CA: SRI Int'l, Artif. Intel. Ctr.)

Nevins, J. L., et al., Report T-575 (Cambridge, MA: C. S. Draper Labs, Inc.).

Nevins, J. L., et al., "Annual Progress Report No. 2 for the Development of Multi-Moded Remote Manipulator Systems," Report C-3901 (Cambridge, MA: C. S. Draper Labs, Inc., Jan. 1973).

Nevins, J. L., et al., "Report on Advanced Automation," Report R-764 (Cambridge, MA: C. S. Draper Labs, Inc., Nov. 1973).

Nevins, J. L., et al., "Exploratory Research in Industrial Modular Assembly," Report R-800 (Cambridge, MA: C. S. Draper Labs, Inc., March 1974).

Nevins, J. L., and Whitney, D., "Exploratory Research in Industrial Modular Assembly - A Status Report," Report P-101 (Cambridge, MA: C. S. Draper Labs, Inc., Aug. 1974), also in Proc. 4th Int'l Symp. on Indust. Robots, Tokyo, 19-21 November, 1977 (Tokyo: Japan Indust. Robot Assoc.).

Nevins, J. L., et al., "Exploratory Research in Industrial Modular Assembly, 1 Feb. 1974 -- 30 Nov. 1974," Report R-850,(Cambridge, MA: C. S. Draper Labs, Inc., Dec. 1974).

Nevins, J. L., and Whitney, D., "Adaptable-Programmable Assembly Systems: An Information and Control Problem," Report P-149

(Cambridge, MA: C. S. Draper Labs, Inc., Feb. 1975), and in Proc. IEEE INTERCON 75, New York, NY, April 1975 (Piscataway, NJ: IEEE Svc. Ctr.).

Nevins, J. L., et al., "Exploratory Research in Industrial Modular Assembly, 1 Dec. 1974 -- 31 Aug. 1975," Report R-921 (Cambridge, MA: C. S. Draper Labs, Inc., Sep. 1975).

Nevins, J. L., "Sensors for Industrial Automation," in McGraw Hill Yearbook, Science and Technology (New York: Mcgraw-Hill, 1975).

Nevins, J. L., et al., "Exploratory Research in Industrial Modular Assembly, 1 Sep. 1975 -- 31 Aug. 1976," Report R-996 (Cambridge, MA: C. S. Draper Labs, Inc., Sep. 1976).

Nevins, J. L., et al., "Exploratory Research in Industrial Modular Assembly, 1 Sep. 1976 -- 31 Aug., 1977," Report R-1111 (Cambridge, MA: C. S. Draper Labs, Inc., Sep. 1977).

Nevins, J. L., et al., "Research Issues for Automatic Assembly," in Proc. 1st IFAC Intl. Symp. on Info. and Control Problems in Mfg. Tech., Tokyo, Japan, Oct. 1977.

Nevins, J. L., and Whitney, D., "Assembly Research and Manipulaton," in Proc. 1977 IEEE Decision and Control Conf., New Orleans, LA, Dec. 1977.

Nevins, J. L., et al., "Research Issues for Automatic Assembly," in Proc. 15th Numerical Control Society Annual Meeting & Technical Conference, Chicago, IL, April 1978.

Nevins, J. L., and Whitney, D., "Computer-controlled Assembly," Sci. Amer., Vol. 238, No. 24 (Feb. 1979), pp. 62 ff.

Niemi, A., Malinen, P., and Koskinen, K., "Digitally Implemented Sensing and Control Functions for a Standard Industrial Robot," Proc. 7th Int'l Symp. on Indust. Robots, Tokyo, 19-21 October, 1977 (Tokyo: Japan Indust. Robot Assoc.), pp. 487-495.

Nitzan, D., Rosen, C. A., et al., "Machine Intelligence Research Applied to Industrial Automation," 9th Report, Nat'l Sci. Found. Grants APR-75-13074 and DAR-78-27-128, SRI Projects 4391 and 8487, (Menlo Park, CA: SRI Int'l, Artif. Intel. Ctr., Aug. 1979).

Norbedo, R. A., "A Structured Software System for Industrial Automation," Proc. 7th Int'l Symp. on Indust. Robots, Tokyo, 19-21 October, 1977 (Tokyo: Japan Indust. Robot Assoc.), pp. 139-144.

Olsztyn, J., "An Application of Computer Vision to a Simulated Assembly Task," Publ. GMR-1483 (Warren, MI: Gen. Motors Res. Labs., GM Corp., Dec. 1973).

Park, W. T. and Burnett, D. L., "An Interactive Incremental Compiler for More Productive Programming of Computer-Controlled Industrial Robots and Flexible Automation Systems," Proc. 9th Int'l Symp. on Indust. Robots, Wash., D.C., 13-15 March, 1979 (Dearborn, MI: Soc. Manuf. Eng.'s, Mkt. Svc. Dept.), pp. 281-298.

Paul, R. L., "Modelling, Trajectory Calculation and Servoing of a Computer Controlled Arm," Memo AIM-177, Report STAN-CS-72-311 (Stanford, CA: Stanford Artif. Intel. Proj., Nov., 1972).

Paul, R. L., et al., "Advanced Industrial Robot Control Systems," First Report, N.S.F. Grant APR-77-14533, Report No. TR-EE 78-25 (West Lafayette, IN: Purdue U., School of E. E., May 1978).

Paul, R. L., et al., "Advanced Industrial Robot Control Systems," Second Report, N.S.F. Grant APR-77-14533, Report No. TR-EE 79-35 (West Lafayette, IN: Purdue U., School of E. E., July 1979).

Prajoux, R., "A Step Toward the Handling of Parts Carried by an Overhead Conveyor: A Robot System using a Fast Vision Sensor to Track a Hanging Object," supported by NSF Grant APR-75-13074, and by C.N.R.S and I.R.I.A. (France) Tech. Note 208 (Menlo Park, CA: SRI Int'l, Artif. Intel. Ctr., Dec. 1979).

Resnick, B., "Robot Interface: Switch Closure and Beyond," Paper MS78-698 (Dearborn, MI: Soc. Manuf. Eng.'s, Mkt. Svc. Dept.), and in Proc. Robots II Conf., Nov. 1977.

Rosen, C. A., et al., "Exploratory Research in Advanced Automation," 1st Report, NSF Grant GI38100X, SRI Project 2591, (Menlo Park, CA: SRI Int'l, Artif. Intel. Ctr., Dec. 1973).

Rosen, C. A., Nitzan, D., et al., "Exploratory Research in Advanced Automation," 2nd Report, NSF Grant GI38100X1, SRI Project 2591, (Menlo Park, CA: SRI Int'l, Artif. Intel. Ctr., Aug. 1974).

Rosen, C. A., Nitzan, D., et al., "Exploratory Research in Advanced Automation," 3rd Report, NSF Grant GI38100X1, SRI Project 2591, (Menlo Park, CA: SRI Int'l, Artif. Intel. Ctr., Dec. 1974).

Rosen, C. A., Nitzan, D., et al., "Exploratory Research in Advanced Automation," 4th Report, NSF Grant GI38100X1, SRI Project 2591, (Menlo Park, CA: SRI Int'l, Artif. Intel. Ctr., June 1975).

Rosen, C. A., Nitzan, D., et al., "Exploratory Research in Advanced Automation," 5th Report, NSF Grant GI38100X1, SRI Project 4391, (Menlo Park, CA: SRI Int'l, Artif. Intel. Ctr., Jan. 1976).

Rosen, C. A., Nitzan, D., et al., "Machine Intelligence Research Applied to Industrial Automation," 6th Report, NSF Grant APR-75-13074, SRI Project 4391, (Menlo Park, CA: SRI Int'l, Artif. Intel. Ctr., Nov. 1976).

Rosen, C. A., Nitzan, D., et al., "Machine Intelligence Research Applied to Industrial Automation," 7th Report, NSF Grant APR-75-13074, SRI Project 4391, (Menlo Park, CA: SRI Int'l, Artif. Intel. Ctr., Aug. 1977).

Rosen, C. A., Nitzan, D., et al., "Machine Intelligence Research Applied to Industrial Automation," 8th Report, NSF Grant APR-75-13074, SRI Project 4391, (Menlo Park, CA: SRI Int'l, Artif. Intel. Ctr., Aug. 1978).

Shimano, B. E., "The Kinematic Design and Force Control of Computer-Controlled Manipulators," Ph. D. Thesis, Stanford U. Mech. Eng. Dept., Memo AIM-313 and Report STAN-CS-78-660, supported by Advanced Res. Proj's Agency Contract MDA903-76-C-0206 and the Nat'l Sci. Found., (Stanford, CA: Stanford Artif. Intel. Proj., March 1978), 134 pp.

Takeyasu, K., et al., "An Approach to the Integrated Intelligent Robot with Multiple Sensory Feedback: Construction and Control Functions," Proc. 7th Int'l Symp. on Indust. Robots, Tokyo, 19-21 October, 1977 (Tokyo: Japan Indust. Robot Assoc.), pp. 523-530.

Taylor, R. H., "Planning and Execution of Straight-Line Manipulator Trajectories," IBM J. Res. Develop., Vol. 23, No. 4 (July 1979), pp. 424-436).

Will, P. M., and Grossman, D. D., "An Experimental System for Computer Controlled Mechanical Assembly," IEEE Trans. Computers, Vol. C-24, No. 98 (Sep. 1979), pp. 879-888.

"Computer Science and Technology, NBS-RIA Robotics Research Workshop, 12-13 July, 1977," Publ. 500-29 (Wash., D.C.: Natl. Bur. Stds., July 1977).

Wang, D., "Comments on NBS Hierarchical Robot Control System," MAT Memo 5271, Aug. 3, 1977.

Whitney, D., "Response for McDonnell Douglas Software Questionaire," MAT Memo 7031, Nov. 15, 1978.

Advanced Vision Research

Agin, G., "Representation and Description of Curved Objects," Memo AIM-173 (Stanford, CA: Stanford Artif. Intel. Proj., 1972).

Agin, G., "Computer Description of Curved Objects," IEEE Trans. Comp., Vol. C-25, No. 4 (April 1976), pp. 439-449.

Eskenazi, R., and Wilf, J., "Low-level Processing for Real-Time Image

Analysis" (to be published), N.A.S.A. Contract NAS7-100 (Pasadena, CA: Jet Propulsion Lab, Cal. Inst. Tech., 1979).

Gara, A., "Real-Time Optical Correlation of 3-Dimensional Scenes," Publ. GMR-2145 (Warren, MI: Gen. Motors Res. Labs., GM Corp., 1976).

O'Handley, D. A., "Scene Analysis in Support of a Mars Rover," Computer Graphics and Image Processing, Vol. 2 (New York: Academic Press, 1973), pp. 281-297.

Yachida, M. and Tsuji, S., "A Machine Vision for Complex Industrial Parts with Learning Capacity," 4th Int'l Jt. Conf. on Artificial Intelligence, Tbilisi, Georgia, U.S.S.R., 3-8 Sep. 1975 (Cambridge, MA: The Artif. Intell. Lab., Publications Dept., 545 Technol. Sq.), pp. 819-825.

Yakimovsky, Y., and Cunningham, R., "A System for Extracting Three-Dimensional Measurements from a Stereo Pair of TV Cameras," Computer Graphics and Image Processing, Vol. 7 (New York: Academic Press, 1978), pp. 195-210.

"Pattern Information Processing System," (Tokyo: Electrotechnical Lab., Agency of Indust. Science & Technol., Min. of Int'l Trade & Industry, 1978).

Application Criteria and New Robotic Applications

Ballard, R. L., "Criteria for Picking an Industrial Robot," Paper MS74-150 (Dearborn, MI: Soc. Manuf. Eng.'s, Mkt. Svc. Dept., 1974).

Blanding, W., "Potential Warehouse Applications of Industrial Robots," The Industrial Robot, Vol. &, No. 3 (Sep. 1979), pp. 125-129.

Brewer, A., et al., "Computer-Based Automation of Discrete Product Manufacture -- a Preliminary Discussion of Feasibility and Impact," (Wash., D.C.: Natl. Tech. Info. Service, U. S. Dept. of Commerce, July 1974).

Dawson, B. L, "Moving Line Applications with a Computer Controlled Robot," SME paper MS77-742 (Dearborn, MI: Soc. Manuf. Eng.'s, Mkt. Svc. Dept., 1977), also in Proc. Robots II Conf., Nov. 1977.

Elkins, S., "The Robot-Driven Waterjet Cutter," Robotics Today (Winter 1979-80), pp. 24-25.

Kuehn, R., Jr., "Requirements of Robotics Systems for Airframe Manufacture," Proc. 8th Int'l Symp. on Indust. Robots, Vol. I, Stuttgart, 30 May - 1 June, 1978 (Bedford, England: Int'l. Fluidics Services, Ltd.), pp. 329-336.

Lynch, P. M., "Study of Alternate Programmable Assembly Equipment Designs Using Technological and Economic Models," Final Report, Nat'l Sci. Found. Grant ENG 77-06447, (New Orleans, LA: Tulane U., Dept. Mech. Eng., June 1979).

Macri, G., "Anaylsis of First UTD Installation Failures," Paper MS77-735 (Dearborn, MI: Soc. Manuf. Eng.'s, Mkt. Svc. Dept., 1978).

Mosher, R., "Robotic Painting -- the Automotive Potential," Paper MS77-735 (Dearborn, MI: Soc. Manuf. Eng.'s, Mkt. Svc. Dept., 1977).

Rogers, P. F., "A Time and Motion Method for Industrial Robots," The Industrial Robot, Vol. 5, No. 4 (Dec. 1978) pp. 187-192.

Sheridan, T., "Performance Evaluation of Programmable Robots and Manipulators," Report of a Workshop held at Annapolis, MD, Oct. 23-25, 1975, Publ. 459 (Wash., D.C.: Natl. Bur. Stds.).

Weichbrodt, B., "Some Special Applications for ASEA Robots -- Deburring of Metal Parts in Production," Paper MS77-736 (Dearborn, MI: Soc. Manuf. Eng.'s, Mkt. Svc. Dept., 1977).

"Future Space Programs," Hearings before the Comm. on Science and Technology, U.S. House of Rep., 95th Congress, Second Session, No. 63, Jan. 24-26, 1978.

"Machine Intelligence and Robotics: Report of the NASA Study Group," Executive Summary (Pasadena, CA: Jet Propulsion Lab, Cal. Inst. Tech., Sep. 1979).

"Manufacturing Technology -- A Changing Challenge to Improve Productivity," Report to the Congress by the Comptroller General of the U. S. (June 1976), n. a.

Artificial Intelligence Research on Robots

Ambler, A. P., et al., "A Versatile Computer-Controlled Assembly System," 3RD Int'l Jt. Conf. on Artificial Intelligence, Stanford, CA, 20-23 Aug. 1973 (Menlo Park, CA: SRI Int'l, Artif. Intel. Ctr.) †, pp. 298-307.

Bolles, R. C., "Verification Vision within a Programmable Assembly System," Ph. D. Thesis, Stanford U., Memo AIM-295, STAN-CS-77-591, (Stanford, CA: Stanford Artif. Intel. Proj., Dec. 1976), 245 pp.

Buttler, W. P., "Self-Navigating Robot," Tech. Supp. Package NPO-14190 (Wash. D.C.: Nat'l Aeron. & Space Admin., Technol. Util. Branch), summary in NASA Tech Briefs (Spring 1978), pp. 31-32.

Lozano-Perez, T., "The Design of a Mechanical Assembly System," Master's Thesis, Report No. 397 (Cambridge, MA: MIT Artif. Intel. Lab., 1978).

Popplestone, R., Ambler, A., and Bellos, I., "An Interpreter for a Language for Describing Assemblies," paper submitted to Artificial Intelligence Journal, available as DAI Research Paper No. 125 (Edinburgh, Scotland: U. Edinburgh, Dept. of Artificial Intelligence, Sep. 1979).

Rieger, C., "Artificial Intelligence Programming languages for Computer-Aided Manufacturing," Report TR-595, Office of Naval Res. Contract N00014-76C-0477 (College Park, MD: U. Maryland, Comp. Sci. Dept., Sep. 1977).

Srinivas, S., "Error Recovery in Robot Systems," Ph. D. Thesis (Pasadena, CA: CA Inst. of Tech., Dec. 1977), 118 pp.

Taylor, R. H., "A Synthesis of Manipulation Control Programs from Task-Level Specifications," Ph. D. thesis, Stanford U., Memo 282 (Stanford, CA: Stanford Artif. Intel. Proj., 1976).

Tsuji, S. and Nakamura, A., "Recognition of an Object in a Stack of Industrial Parts," 4th Int'l Jt. Conf. on Artificial Intelligence, Tbilisi, Georgia, U.S.S.R., 3-8 Sep. 1975 (Cambridge, MA: The Artif. Intell. Lab., Publications Dept., 545 Technol. Sq.), pp. 811-818.

Udupa, S., "Collision Detection and Avoidance in Computer-Controlled Manipulators," Ph. D. thesis, C.I.T. (Pasadena, CA: CA Inst. of Tech., 1977), 33 pp.

Attitude of Unions towards Robotization

Weekley, T. L., "A View of the United Automobile, Aerospace and Agricultural Implement Workers of America (UAW) Stand on Indutrial Robots," in Proc. Robots II Conf., Nov. 1977.

Weekley, T. L., "The UAW Speaks Out on Industrial Robots," Robotics Today (Winter 1979-80), pp. 25-27.

Compliance and Accommodation Technology

Drake, S., "Using Compliance in Lieu of Sensory Feedback for Automatic Assembly," Report T-657 (Cambridge, MA: C. S. Draper Labs, Inc., Sep. 1977).

Drake, S., et al., "High Speed Robot Assembly of Precision Parts Using Compliance Instead of Sensory Feedback," Proc. 7th Int'l Symp. on Indust. Robots, Tokyo, 19-21 October, 1977 (Tokyo: Japan Indust. Robot Assoc.), pp. 87-98.

Rebman, J., "Compliance: The Forgiving Factor," Robotics Today (Fall 1979), pp. 29-34.

Selvage, C., "Assembly of Interference Fits by Impact and Constant Force Methods," M. Sc. Thesis, M.I.T., and Report T-693 (Cambridge, MA: C. S. Draper Labs, Inc., June 1979).

Simunovic, S. N., "An Information Approach to Parts Mating," Sc. D. Thesis in Mech. Eng., M. I. T., Report T-690. (Cambridge, MA: C. S. Draper Labs, Inc., April 1979).

Watson, P., "A Multidimensional System Analysis of the Assembly Process as Performed by a Manipulator," Paper MR76-613 (Dearborn, MI: Soc. Manuf. Eng.'s, Mkt. Svc. Dept., 1976), 16 pp.

Whitney, D., "What is the Remote Center Compliance (RCC) and What Can It Do?" Report P-728 (Cambridge, MA: C. S. Draper Labs, Inc., Nov. 1978).

Computer Graphics for Simulation of Robotic Operations

Baumgart, B. G., "GEOMED -- A Geometric Editor," Memo AIM-232 and Report STAC-CS-74-414 (Stanford, CA: Stanford Artif. Intel. Proj., May 1974).

Bonney, M. C., et al., "Using SAMMIE for Computer Aided Workplace and Work Task Design," Proc. 6th Congress, Intl. Ergonomics Assoc. (MD: U. Maryland, 1976).

Foley, J. D., and Pizer, S. M, "Review of Graphic Languages," Computer Graphics and Image Processing, Vol. 1, (New York, NY: Academic Press, 1973), pp. 196-201.

Hall, W., Jervis, B., and Jervis, J., "GLISP - A LISP Based Graphic Language," (Canada: U. British Columbia, Dept. Comp. Sc., 1973).

Heginbotham, W. B., Dooner, M., and Kennedy, D., "Analysis of Industrial Robot Behaviours by Computer Graphics," in Proc. of 3rd CISM-IFTOMM Symp. on Theory and Practice of Robots and Manipulators, Udine, Italy, 12-15 Sep. 1978, (New York: Springer-Verlag, 1978).

Heginbotham, W. B., Dooner, M., and Kennedy, D. N., "Computer Graphics Simulation of Industrial Robot Interaction," Proc. 7th Int'l Symp. on Indust. Robots, Tokyo, 19-21 October, 1977 (Tokyo: Japan Indust. Robot Assoc.), pp. 169-176.

Heginbotham, W. B., Dooner, M., and Kennedy, D. N., "Rapid Assessment of Industrial Robots Performance by Interactive Computer Graphics," Proc. 9th Int'l Symp. on Indust. Robots, Wash., D.C., 13-15 March, 1979 (Dearborn, MI: Soc. Manuf. Eng.'s, Mkt. Svc. Dept.), pp. 563-574.

Heginbotham, W. B., Dooner, M., and Case, K., "Assessing Robot Performance with Interactive Computer Graphics," Robotics Today (Winter 1979-80), pp. 33-35.

Newman, W. M., and Sproul, R. F., Principles of Interactive Computer Graphics, (New York, NY: McGraw-Hill, 1973).

Williams, R. and Krammer, G., "EX.GRAF: An Extensible Language Including Graphical Operations," Computer Graphics and Image Processing, Air Force Office of Sci. Res. Grant AF-AFOSR-70-1854 (New York: Academic Press, 1972, pp. 317-340).

Current Practice and Commercial Systems for Industrial Vision

Baird, M., "Relational Object Models for Object Location," Res. Publ. GMR-1984 (Warren, MI: Gen. Motors Res. Labs., GM Corp., Oct. 1975).

Brain, A. E., "Lenses for Industrial Automation -- Part One: A Brief Review of Basic Optics," Tech. Note 201 (Menlo Park, CA: SRI Int'l, Artif. Intel. Ctr., Nov. 1979).

Branaman, L. A., "Optomation (TM) Instrument System -- Electronic Vision for Process Control" (describes G. E.'s commercial product), Paper AD78-753 (Dearborn, MI: Soc. Manuf. Eng.'s, Mkt. Svc. Dept., 1978), 11 pp.

Gregory, P., and Alexandre, N. H., "A Software-Based Television Image-Analysis System," (descripion of Joyce-Loebl's MAGISCAN product), Society of Motion Picture and Television Engineers Journal, Vol. 88 (Feb. 1979), pp. 117-118.

"Camera Zooms in on Auto Focusing," New Scientist, Vol. 83, n. no. (June 8, 1978), p. 672.

End Effectors, Robot Accessories and Actuator Technology

Drazen, P. J. and Jeffery, M. F., "Some Aspects of an Electro-Pneumatic Industrial Manipulator," Proc. 8th Int'l Symp. on Indust. Robots, Vol. I, Stuttgart, 30 May - 1 June, 1978 (Bedford, England: Int'l. Fluidics Services, Ltd.), pp. 396-405.

Griffith, J. E., et al., "Quasi-Liquid Vise for a Computer-Controlled Manipulator," Report RC 5451, . 23797 (Yorktown Heights, NY: IBM T. J. Watson Res. Ctr., June 1975).

Hill, J. W. and Sword A. .J., , "Programmable Part Presenter Based on Computer Vision and Controlled Tumbling," supported by Nat. Sci. Found Grants APR-75-13074 and 15 industrial affiliates, SRI Projects 4391 and 6284, Tech. Note 194 (Menlo Park, CA: SRI Int'l, Artif. Intel. Ctr., March 1980), 12 pp.

Hirose, S. and Umetani, Y., "The Development of Soft Gripper for the Versatile Robot Hand," Proc. 7th Int'l Symp. on Indust. Robots, Tokyo, 19-21 October, 1977 (Tokyo: Japan Indust. Robot Assoc.), pp. 353-360.

Okada, T., "On a Versatile Finger System," Proc. 7th Int'l Symp. on Indust. Robots, Tokyo, 19-21 October, 1977 (Tokyo: Japan Indust. Robot Assoc.), pp. 345-352.

Rovetta, A. and Casarico, G., "On the Prehension of a Robot Mechanical Hand: Theoretical Analysis and Experimental Tests," Proc. 8th Int'l Symp. on Indust. Robots, Vol. I, Stuttgart, 30 May - 1 June, 1978 (Bedford, England: Int'l. Fluidics Services, Ltd.), pp. 444-451.

Skinner, F., "Design of a Multiple Prehension Manipulator System," ASME Publ. No. 74-DET-25 (Oct. 1974).

Skinner, F., "Multiple Prehension Hands for Assembly Robots," Proc. 5th Int'l Symp. on Indust. Robots, Chicago, IL, 22-24 September, 1975 (Chicago, IL: IIT Research Inst.), PP. 77-88.

"Rigid Coupling is Also Flexible," Tech. Supp. Package MSC-16488 (Wash. D.C.: Nat'l Aeron. & Space Admin., Technol. Util. Branch), summary in NASA Tech Briefs (Spring 1978), p. 105.

"Dyna-Slide Outfeed Conveyor," (product brochure, brush tables & conveyors) Bulletin 744-1-15M (E. Syracuse, NY: Dyna-Slide, Inc., 1979).

"Controlled Linear Induction Actuators," (U.K. Patent Application

28376/74) NRDC Inventions (London, England: Nat'l Res. & Devel. Corp., May 1977).

Industrial Automation Surveys and Research Summaries

Abraham, R., Stewart, R., and Shum, L., "The State-of-the-Art in Tactile and Force Sensing," in State-of-the-Art in Adaptable Programmable Assembly Systems (Bedford, England: Int'l. Fluidics Services, Ltd., 1977), pp. 86-96.

Abraham, R., Stewart, R., and Shum, L., "State-of-the-Art in Vision Systems," in State-of-the-Art in Adaptable Programmable Assembly Systems (Bedford, England: Int'l. Fluidics Services, Ltd., 1977), pp. 97-120.

Abraham, R., Stewart, R., and Shum, L., "The State-of-the-Art in Computer Hardware and Software Systems," in State-of-the-Art in Adaptable Programmable Assembly Systems (Bedford, England: Int'l. Fluidics Services, Ltd., 1977), pp. 121-144.

Abraham, R., Stewart, R., and Shum, L., "State of the Art in Adaptable-Programmable Assembly Systems," Nat'l Sci. Found. Grant ISP-76-24164 (Pittsburgh, PA: Westinghouse R&D Center, May 1977).

Agin, G., "Vision Systems for Inspection and Manipulator Control," Proc. 1977 Jt. Automatic. Control Conf., San Francisco, CA, 22-24 June, 1977 (Piscataway, NJ: IEEE Svc. Ctr.), pp. 132-138.

Albus, J. S., and Evans, J., "Robot Systems," Sci. Am., Vol. 234, No. 2 (Feb. 1976), pp. 77 ff.

Bowman, D., "Industrial Robots -- A Practical Handbook," (Monroeville, PA.: Intl. Material Mgt. Soc., 1976).

Engelberger, J. F., "Robotics: 1984," Robotics Today (Fall 1979), pp. 26-27.

Eversheim, W., et al., "Fields of Research," (Aachen, W. Germany: Laboratorium fur Werkzeugmaschinen und Betriebslehre, Technische Hochschule Aachen, Jan. 1978).

Harrington, J., Computer Integrated Manufacturing (New York, Industrial Press, Inc., 1973).

Kirk, F., and Rabowi, R., Instrumentation (Chicago, IL: Amer. Tech. Soc. Press).

Kirk, F., Basic Industrial Measurement and Control (New York, NY: Penton Publishing Co., Ed. Div., 1973).

Munson, G. E., "Robot Control--An Overview," Proc. 1977 Jt. Automatic. Control Conf., San Francisco, CA, 22-24 June, 1977 (Piscataway, NJ: IEEE Svc. Ctr.), pp. 714-719.

Nitzan, D. and Rosen, C. A. "Programmable Industrial Automation," NSF Grant GI-38100X1, Tech. Note 133 (Menlo Park, CA: SRI Int'l, Artif. Intel. Ctr., Feb. 1979).

Nitzan, D., "Flexible Automation Program at SRI" Proc. 1979 Jt. Automatic. Control Conf., Denver, CO, 17-21 June, 1979, L. C. cat. no. 79-52918, (New York, NY: Amer. Inst. of Chem. Eng's), pp. 754-759, also available from Menlo Park, CA: SRI Int'l, Artif. Intel. Ctr.

Nitzan, D., "Robotic Automation Program at SRI," Proc., MIDCON/79, Chicago, IL, 6-8 Nov. 1979, also available from Menlo Park, CA: SRI Int'l, Artif. Intel. Ctr..

Olsztyn, J., "The Machine Perception Project at the GM Research Laboratories," Publ. GMR-1979 (Warren, MI: Gen. Motors Res. Labs., GM Corp., Sep. 1975).

Park, W. T., "Minicomputer Software Organization for Control of Industrial Robots," Proc. 1977 Jt. Automatic. Control Conf., San Francisco, CA, 22-24 June, 1977 (Piscataway, NJ: IEEE Svc. Ctr.), pp. 164-171.

Park, W. T., "Robotics Research Trends," Tech. Note 160, (Menlo Park, CA: SRI Int'l, Artif. Intel. Ctr., March 1978).

Rooks, B. (ed.), Industrial Robots -- A Survey (Bedford, England: Int'l. Fluidics Services, Ltd., 1972).

Rosen, C. A., and Binford, T. O., "Industrial Systems," Workshop Report, Proc. Research Workshop on Sensors for Automation, NSF Grant GK-37898 (Cambridge, MA: C. S. Draper Labs, Inc., April 1973), pp. 21-26.

Schreiber, R., "Unimation Inc.'s PUMA Robot Arm System," Robotics Today (Winter 1979-80), pp. 42-43.

Shapiro, S. F., "Vision Expands Robotic Skills for Industrial Applications," Computer Design, Vol. 18, No. 9 (Sep. 1979), pp. 78-87).

Smith, B., "NBS Program in Automation Technology," internal memo (Wash., D.C.: Natl. Bur. Stds., Jan. 1979).

Sowen, D., III and Vacroux, A., "Microcomputers -- An Introduction to Microcomputers and their Applications," ???.

Tanner, W. R., Industrial Robots, Volume I, Fundamentals, and Volume II, Applications (Dearborn, MI: Soc. Manuf. Eng.'s, Mkt. Svc. Dept., 1979).

Warnecke, H., and Schraft, R., "Government-sponsored Research in the Field of Industrial Robots in the Federal Republic of Germany," The Industrial Robot, Vol. 4, No. 1 (March 1977).

Weaver, J. A., "Smart Sensors are Enhancing Flexible Automation," Assembly Engineering, Vol. 22, No. 8 (Aug. 1979), pp. 32-33.

Weisel, W., "What can Medium Technology Robots Do?" Paper MS77-737 (Dearborn, MI: Soc. Manuf. Eng.'s, Mkt. Svc. Dept., 1977).

Wisnosky, D., "Worldwide Computer-Aided Manufacturing Survey, 13 Aug. -- 10 Sep. 1977," (Integrated Comp. Aided Manufact. Program, Air Force Materials Lab., Wright-Patterson AFB, OH, Dec. 1977).

Young, John F., Robotics (New York, John Wiley & Sons, 1973).

"Robots in Metal Working," American Machinist, Vol. 119, No. 20 (Nov. 1975), p. 87.

Industrial Vision Research

Agin, G., "An Experimental Vision System for Industrial Application," Tech. Note 103 (Menlo Park, CA: SRI Int'l, Artif. Intel. Ctr., June 1975).

Agin, G., and Duda, R., "SRI Vision Research for Advanced Industrial Automation," Proc. 2nd USA-Japan Computer Conf., Tokyo, Japan, (Menlo Park, CA: SRI Int'l, Artif. Intel. Ctr., Aug. 1975).

Agin, G., "Servoing with Visual Feedback," Technical Note 149 (Menlo Park, CA: SRI Int'l, Artif. Intel. Ctr., July 1977).

Agin, G., "Real Time Control of a Robot with a Mobile Camera," Tech. Note 179 (Menlo Park, CA: SRI Int'l, Artif. Intel. Ctr., Feb. 1979).

Anuashvili, A. N. and Zotov, V. D., "A Simple Real Time Visual System for an Industrial Robot," Proc. 7th Int'l Symp. on Indust. Robots, Tokyo, 19-21 October, 1977 (Tokyo: Japan Indust. Robot Assoc.), pp. 507-514.

Dodd, G. G., and Rossol, L. (eds.), Computer Vision and Sensor-Based Robots (New York: Plenum Press, 1979).

Gleason, G., and Agin, G., "A Modular Vision System for Sensor-

Controlled Manipulation and Inspection," Proc. 9th Int'l Symp. on Indust. Robots, Wash., D.C., 13-15 March, 1979 (Dearborn, MI: Soc. Manuf. Eng.'s, Mkt. Svc. Dept.), pp. 57-70, and Tech. Note 178 (Menlo Park, CA: SRI Int'l, Artif. Intel. Ctr., Feb. 1979).

Holland, S., "An Approach to Programmable Computer Vision for Robotics," Paper MS77-747 (Dearborn, MI: Soc. Manuf. Eng.'s, Mkt. Svc. Dept., 1977).

Holland, S., "A Programmable Computer Vision System based on Spatial Relationships," Publ. GMR-2078 (Warren, MI: Gen. Motors Res. Labs., GM Corp., Feb. 1976).

Kelley, R., "Workplace Transportation by Robots Usng Vision," Paper MS77-746 (Dearborn, MI: Soc. Manuf. Eng.'s, Mkt. Svc. Dept., 1977).

Kelley, R. and Silvestro, K., "V/I, A Visual Instruction Software System for Programming Industrial Robots," The Industrial Robot, Vol. 4, No. 2 (June 1977), pp. 59-75.

Perkins, W. A., "Multilevel Vision Recognition System," Publ. GMR-2125 (Warren, MI: Gen. Motors Res. Labs., GM Corp., April 1976).

Perkins, W. A., "A Model-Based Vision System for Industrial Parts," I.E.E.E. Transactions on Computers, Vol. C-27 (Feb. 1978), pp. 126-143.

Perkins, W. A., "Computer Vision Classification of Automotive Control Arm Bushings," Proc. Computer Software & Applications Conf., Chicago Il., Nov. 6-8, 1979, pp. 344-349.

Spur, G., Kraft, H. R., and Sinninng, H., "Computer Controlled Object Identification Using Visual Sensors," Proc. 8th Int'l Symp. on Indust. Robots, Vol. I, Stuttgart, 30 May - 1 June, 1978 (Bedford, England: Int'l. Fluidics Services, Ltd.), pp. 155-164.

"Grid Circle Analysis of Stamped Metals can be Automated," summary of a General Motors Res. Lab. report, Production Engineering, Vol. 26, No. 4 (April 1979), p. 14.

VanderBrug, G. J., Albus, J. S., and Barkmeyer, E., "A Vision System for Real Time Control of Robots," Proc. 9th Int'l Symp. on Indust. Robots, Wash., D.C., 13-15 March, 1979 (Dearborn, MI: Soc. Manuf. Eng.'s, Mkt. Svc. Dept.), pp. 213-232.

Vanderbrug, G. J., Albus, J. S., and Barkmeyer, E., "A Vision System for Real-Time Robot Control," Robotics Today (Winter 1979-80), pp. 20-22.

"Low-intensity X-Ray and Gamma-Ray Imaging Device," Tech. Supp. Package NPO-14073 (Wash. D.C.: Nat'l Aeron. & Space Admin., Technol. Util. Branch), summary in NASA Tech Briefs (Spring 1978), pp. 67-68.

Manipulator Control Systems and Techniques

Albus, J. S., "A New Approach to Manipulator Control: The Cerebellar Model Articulation Controller (CMAC)," J. Dynamic Systems, Measurement and Control, Trans. of the A.S.M.E. (Sep. 1975), pp. 220-227.

Albus, J. S., "Data Storage in the Cerebellar Model Articulation Controller (CMAC)," J. Dynamic Systems, Measurement and Control, Trans. of the A.S.M.E. (Sep. 1975), pp. 228-233.

Blanchard, M., "Digital Control of a Six-Axis Manipulator," Working Paper 129 (Cambridge, MA: MIT Artif. Intel. Lab., 1976).

Coiffet, P., et al., "Real Time Problems in Computer Control of Robots," Proc. 7th Int'l Symp. on Indust. Robots, Tokyo, 19-21 October, 1977 (Tokyo: Japan Indust. Robot Assoc.), pp. 145-152.

Corwin, M., "The Benefits of a Computer Controlled Robot," Paper MS75-273 (Dearborn, MI: Soc. Manuf. Eng.'s, Mkt. Svc. Dept.), and Proc. 5th Int'l Symp. on Indust. Robots, Chicago, IL, 22-24 September, 1975 (Chicago, IL: IIT Research Inst.), pp. 453-470.

Craig, J. J., and Raibert, M. H., "A systematic Method of Hybrid Position/Force Control of a Manipulator," Proc. Computer Software & Applications Conf., Chicago Il., 6-8 Nov. 1979, pp. 446-451.

Cunningham, C., "Robot Flexibility Through Software," Proc. 9th Int'l Symp. on Indust. Robots, Wash., D.C., 13-15 March, 1979 (Dearborn, MI: Soc. Manuf. Eng.'s, Mkt. Svc. Dept.), pp. 297-308.

Dobrotin, B. and Lewis, R. A., "A Practical Manipulator System," 5th Int'l Jt. Conf. on Artificial Intelligence, Cambridge, MA, 22-25 Aug., 1977 (Pittsburgh, PA: Carnegie-Mellon U., Dept. of Comp. Sci.), pp. 723-732.

Doty, K. L., "A Distributed Microcomputer Network for Controlling a Robotic Manipulator," Proc. NSF Workshop on the Impact on the Academic Community of Required Research Activity for Generalized Robotic Manipulators, Gainesville, FL, 8-10 Feb., 1978 (Gainesville, FL: M.E. Dept., U. Florida), pp. 233-250.

Duffy, J. and Martinovic, R., "Kinematics of 3R-3P Computer controlled Manipulators," Proc. 7th Int'l Symp. on Indust. Robots, Tokyo, 19-21 October, 1977 (Tokyo: Japan Indust. Robot Assoc.), pp. 67-78.

Fournier, A., and Khalil, W., "Co-ordination and Reconfiguration of Mechanincal Redundant Systems," Proc. Int'l Conf. on Cyb. & Soc.,

Wash., D. C., 19-21 Sep., 1977 (Piscataway, NJ: IEEE Svc. Ctr.), pp. 227-231.

Gill, A., Paul, R. L., and Scheinman, V., "Computer Manipulator Control, Visual Feedback and Related Problems," (Proc. of 1st CISM-IFTOMM Symp. on Theory and Practice of Robots and Manipulators, Vol. II, Udine, Italy, 5-8 Sep. 1973, ISBN 3-211-81252-0, 0-387-81252-0 Courses & Lectures No. 201, Int'l Ctr. for Mech. Sciences, (New York: Springer-Verlag, 1974)), pp. 31-50.

Hocken, J. A., et al., "Three Dimensional Metrology," in Annals of the C.I.R.P., Vol. 26, 1977, reprints available from main author (Wash., D.C.: Natl. Bur. Stds.).

Hohn, R. E., "Application Flexibility of a Computer-Controlled Industrial Robot," Paper MR76-603 (Dearborn, MI: Soc. Manuf. Eng.'s, Mkt. Svc. Dept., 1976) also in Proc. 1st North American Indust. Conf., 1976.

Hohn, R. E., "Computed Path Control for an Industrial Robot," Proc. 8th Int'l Symp. on Indust. Robots, Vol. I, Stuttgart, 30 May - 1 June, 1978 (Bedford, England: Int'l. Fluidics Services, Ltd.), pp. 327-337.

Holt, H. R., "Computer Robot Controls," Mechanical Engineering, Vol. 99, No. 3 (March 1977), p. 82.

Holt, H. R., "Robot Decision Making," Paper MS77-751 (Dearborn, MI: Soc. Manuf. Eng.'s, Mkt. Svc. Dept., 1977), also in Proc. Robots II Conf., Nov. 1977.

Holt, H. R., "Trends in Robot Controls," Agricultural Engineering, Vol. 59, No. 4 (April 1978), pp. 42-43.

Luh, J. Y. S., and Lin, C. S., "Multiprocessor-Controllers for Mechanical Manipulators," Proc. Computer Software & Applications Conf., Chicago, IL., 6-8 Nov. 1979, pp. 458-463.

Luh, J. Y. S., Walker, M., and Paul, R. L., "On-Line Computational Scheme for Mechanical Manpulators," (to be published), Nat'l Sci. Found. Grants ENG 76-18567 and APR-77-14533 (West Lafayette, IN: Purdue U., School of E. E., 1979).

Makino, H., "A Kinematical Classification of Robot Manipulator," in Proc. 3rd Conf. on Indust. Robot Technol., Tokyo, Japan, Nov. 1977 (Bedford, England: Int'l. Fluidics Services, Ltd.) and Proc. 6th Int'l Symp. on Indust. Robots, Nottingham, England, March 1976.

Mazza, G., Sarzana, F., and Somalvico, M., "Design and Construction of a Microcomputer for an Assembly Robot," The Industrial Robot, Vol. 6, No. 1 (March 1979), pp. 9-14.

McGhee, R. B., "Dynamics and Control of Manipulators and Robotic Systems," Proc. NSF Workshop on the Impact on the Academic Community of Required Research Activity for Generalized Robotic Manipulators, Gainesville, FL, 8-10 Feb., 1978 (Gainesville, FL: M.E. Dept., U. Florida), pp. 251-255.

Paul, R. L., "WAVE - A Model-based Language for Manipulator Control," The Industrial Robot, Vol 4, No. 1 (March, 1977), pp. 10-17.

Paul, R. L., "Robot Software and Servoing," Proc. NSF Workshop on the Impact on the Academic Community of Required Research Activity for Generalized Robotic Manipulators, Gainesville, FL, 8-10 Feb., 1978 (Gainesville, FL: M.E. Dept., U. Florida), pp. 255-259.

Peiper, D., "The Kinematics of Manipulators under Computer Control," Ph. D. Thesis, A.R.P.A. Order No. 457, Contract SD-183, Report CS-116 or Memo AI-72 (Stanford, CA: Stanford Artif. Intel. Proj., Oct. 1968).

Raibert, M., "A State Space Model for Sensorimotor Control and Learning," Memo 351, Nat'l Inst. of Health Grant NIH-5-T01-6MO-1064-14, Office of Naval Res. Contract N00014-75-C-0634 (Cambridge, MA: MIT Artif. Intel. Lab., Jan. 1976).

Roderick, M. D., "Discrete Control of a Robot Arm," Eng. Thesis in E. E., Memo AIM-287, STAN-CS-76-571 (Stanford, CA: Stanford Artif. Intel. Proj., Aug. 1976), 100 pp.

Ruoff, C. F., "PACS - An Advanced Multitasking Robot System," to be published in The Industrial Robot, also internal memo (Pasadena, CA: Jet Propulsion Lab, Cal. Inst. Tech., Robotics and Teleoperators Group, 1979).

Ruoff, C. F., "Teach - A Concurrent Robot Control Language," Proc. Computer Software & Applications Conf., Chicago IL., 6-8 Nov. 1979, pp. 442-445.

Shaket, E., and Freedy, A., "A Model of Man/Machine Communication in Computer Aided Manipulator," Proc. Int'l Conf. on Cyb. & Soc., Wash., D. C., 19-21 Sep., 1977 (Piscataway, NJ: IEEE Svc. Ctr.), p. 773 (1977).

Shimano, B., "VAL: A Versatile Robot Programming and Control System," Proc., Computer Software and Applictions Conference, Chicago, IL., 6-8 Nov. 1979, pp. 878-883.

Snyder, W., "Computer Control of Robots--A Servo Survey," MS76-617 (Dearborn, MI: Soc. Manuf. Eng.'s, Mkt. Svc. Dept., 1976).

Sword, A. J., and Park, W. T., "Location and Acquisition of Objects in Unpredictable Locations," voice control of a Unimate, Tech. Note 102 (Menlo Park, CA: SRI Int'l, Artif. Intel. Ctr., May 1975).

Takase, K., Paul, R. L., and Berg, E. J., "A Structured Approach to Robot Programming and Teaching," Proc. Computer Software & Applications Conf., Chicago IL., 6-8 Nov. 1979, pp. 452-457.

Utkin, V., "Equations of Sliding Mode in Discontinuous Systems" Automation and Remote Control, Part I (1971), pp. 1897-1907, and Part II (1972), pp. 211-219.

Waters, R., "Mechanical Arm Control," Vision Flash 42 (internal paper), Office of Naval Res. Contract No. N00014-70-A-0362-005 (Cambridge, MA: MIT Artif. Intel. Lab., March 1973).

Waters, R., "A Mechanical Arm Control System," Office of Naval Res. Contract No. N00014-70-A-0362-0005, Memo 301 (Cambridge, MA: MIT Artif. Intel. Lab., Jan. 1974).

Whitney, D., "Resolved Motion Rate Control of Manipulators and Human Prostheses," I.E.E.E. Trans. Man-Machine Systems, Vol. MMS-10, No. 2 (June 1969), pp. 47-53.

Williams, R. J., "Dynamic Analysis of the Musculoskeletal System and Robots," (Proc. NSF Workshop on the Impact on the Academic Community of Required Research Activity for Generalized Robotic Manipulators, Gainesville, FL, 8-10 Feb., 1978 (Gainesville, FL: M.E. Dept., U. Florida), pp. 344-349.

Young, K., "Controller Design for a Manipulator Using Theory of Variable Structure Systems," I.E.E.E. Transactions on Systems, Man, and Cybernetics, Vol. SMC-8, No. 2 (Feb. 1978), pp. 101-109.

"Computer Interface for Mechanical Arm," Tech. Supp. Package MFS-23849 (Wash. D.C.: Nat'l Aeron. & Space Admin., Technol. Util. Branch), summary in NASA Tech Briefs (Spring 1978), p. 20.

"Adaptive Control for Weld Skate," Tech. Supp. Package MFS-23620 (Wash. D.C.: Nat'l Aeron. & Space Admin., Technol. Util. Branch), summary in NASA Tech Briefs (Spring 1977), p. 142.

"Branching Capabilities of the T3," Robotics Today (Winter 1979-80), n. a., p. 32.

"NC Cylinder Positions Heavy Loads at High Speed," (product brochure: stepper-motor-controlled hydraulic actuator) (Cambridge, Ontario: Mimik Ltd.), and in Machine Design, 21 July 1977, p. 40.

"Pneumatic Servomechanisms Dynamic Analysis Program," Tech. Supp. Package MFS-23295 (Wash. D.C.: Nat'l Aeron. & Space Admin., Technol. Util. Branch), summary in NASA Tech Briefs (Spring 1978), p. 146.

Manipulator Design

Boykin, W. H., "Position Paper on Robotic Manipulator Systems and Concepts," Proc. NSF Workshop on the Impact on the Academic Community of Required Research Activity for Generalized Robotic Manipulators, Gainesville, FL, 8-10 Feb., 1978 (Gainesville, FL: M.E. Dept., U. Florida), pp. 290-302.

Davies, B. L., and Ihnatowicz, E., "A Three-Degree-of-Freedom Robotic Manipulator," in Proc. Robots IV Conf., Detroit, MI, 30 Oct. - 1 Nov. 1979.

Davies, B. L., and Ihnatowicz, E., "A Three-Degree-of-Freedom Robotic Manipulator," Robotics Today (Winter 1979-80), pp. 28-29.

Dunne, M. J., "An Advanced Assembly Robot," (Unimate 6000 series dual manipulator) Paper MS77-755 (Dearborn, MI: Soc. Manuf. Eng.'s, Mkt. Svc. Dept., 1977).

Flateau, C. R., "Design Outline for Mini-Arms Based on Manipulator Technology," ONR Contract N00014-70-A-0362-005, Memo AI-300 (Cambridge, MA: MIT Artif. Intel. Lab., May 1973).

Nevins, J. L., et al., "A Scientific Approach to the Design of Computer-Controlled Manipulators," Report R-837 (Cambridge, MA: C. S. Draper Labs, Inc., Aug. 1974).

Roth, B., Rastegar, J., and Scheinman, V., "On the Design of Computer Controlled Manipulators," Proc. of 1st CISM-IFTOMM Symp. on Theory and Practice. of Robots and Manipulators, Vol. I, Udine, Italy, 5-8 Sep. 1973, ISBN 3-211-81252-0, 0-387-81252-0 Courses & Lectures No. 201, Int'l Ctr. for Mech. Sciences, (New York: Springer-Verlag, 1974) pp. 93-114.

Scheinman, V., "Design of a Computer-Controlled Manipulator," Adv. Res. Proj. Agency Contract SD-183, Mech. Eng. Thesis, Stanford U., and Memo AI-92 (Stanford, CA: Stanford Artif. Intel. Proj., June 1969).

Shimano, B., and Roth, B., "Dimensional Synthesis of Manipulators," in Proc. of 3rd CISM-IFTOMM Symp. on Theory and Practice of Robots and Manipulators, Udine, Italy, 12-15 Sep. 1978, (New York: Springer-Verlag, 1978).

Stackhouse, T., "A New Concept in Robot Wrist Flexibility," Proc. 9th Int'l Symp. on Indust. Robots, Wash., D.C., 13-15 March, 1979 (Dearborn, MI: Soc. Manuf. Eng.'s, Mkt. Svc. Dept.), pp. 589-600.

Vertut, J., and Liefeois, A., "General Design Criteria of Manipulators," in Proc. of 3rd CISM-IFTOMM Symp. on Theory and Practice of Robots and Manipulators, Udine, Italy, 12-15 Sep. 1978, (New York: Springer-Verlag, 1978).

Ward, M., "Specifications for a Computer Controlled Manipulator," Publ. GMR-2066 (Warren, MI: Gen. Motors Res. Labs., GM Corp., Feb. 1976).

Modern Robotic Practice

Blean, R. B. and Gleason, G., "Computer-assisted Manipulator Training," 1977 IEEE Conf. on Decis. and Control, New Orleans, LA, Dec. 6-9, 1977 (Menlo Park, CA: SRI Int'l, Artif. Intel. Ctr.).

Bollinger, J. G., and Ramsey, P. W., "Computer Controlled Self Programming Welding Machine," in Welding Journal (May, 1979).

Corwin, M., "A Computer Controlled Robot for Automotive Manufacturing," in Proc. Intl. Symp. on Automotive Technology and Automation (Sep. 1977).

d'Auria, A., and Salmon, M., "SIGMA - An Integrated General-Purpose System for Automatic Manipulation," Proc. 5th Int'l Symp. on Indust. Robots, Chicago, IL, 22-24 September, 1975 (Chicago, IL: IIT Research Inst.), pp. 185-202.

Dawson, B. L, "A Computerized Robot Joins the Ranks of Advanced Manufacturing Technology," in Proc. Numerical Control Soc.'s 13th Annual Mtg. and Tech. Conf. (Numerical Control Soc., 1976).

Hasegawa, K., et al., "Programming and Teaching Method for Industrial Robot," Proc. 4th Int'l Symp. on Indust. Robots, Tokyo, 19-21 November, 1977 (Tokyo: Japan Indust. Robot Assoc.), pp. 301-310.

Holmes, J. G., "An Automated Robot Machining System," Proc. 9th Int'l Symp. on Indust. Robots, Wash., D.C., 13-15 March, 1979 (Dearborn, MI: Soc. Manuf. Eng.'s, Mkt. Svc. Dept.), pp. 39-56.

Inagaki, S., "An Easy Programming Controller for Industrial Robots," Proc. 7th Int'l Symp. on Indust. Robots, Tokyo, 19-21 October, 1977 (Tokyo: Japan Indust. Robot Assoc.), pp. 153-160.

Kuzmierski, T., "Robot Development for Aerospace Batch Manufacturing," Proc. 1977 Jt. Automatic. Control Conf., San Francisco, CA, 22-24 June, 1977 (Piscataway, NJ: IEEE Svc. Ctr.), pp. 704-709.

Lockett, J. H., "Small-Batch Production of Aircraft Access Doors Using an Industrial Robot," in Proc. Robots IV Conf., Detroit, MI, 30 Oct. - 1 Nov. 1979.

Lockett, J. H., "The Robotic Work Station in Small-Batch Production," Robotics Today (Winter 1979-80), pp. 17-19.

Okamoto, K., et al., "Grinding Robot on Crooked Surface," Proc. 7th Int'l Symp. on Indust. Robots, Tokyo, 19-21 October, 1977 (Tokyo: Japan Indust. Robot Assoc.), pp. 615-622.

Salmon, M., "Assembly by Robots," The Industrial Robot, Vol. 4, No. 2 (June, 1977), pp. 81-85.

Programming Languages and Software

Dahl, O. J., Dijkstra, E. W. and Hoare, C. A. R., "Structured Programming," (New York, NY: Academic Press, 1972).

Feldman, J. A., "Programming Languages," Sci. Am., Vol. 241, No. 6 (Dec. 1979), pp. 94-112.

Mujtaba, S. and Goldman, R., "AL Users' Manual," Memo AIM-323 and Report STAN-CS-79-718, Nat'l Sci. Found. Grants NSF-APR-74-01390-A04 and NSF-DAR-78-15914, (Stanford, CA: Stanford Artif. Intel. Proj., Jan. 1979), 130 pp.

Requicha, A. A. G., Samuel, N. M. and Voelcker, H. B., "Part and Assembly Description Languages -- II," Memo TM-20a (Rochester, NY: U. Rochester, Production Automation Proj., Nov. 1974).

Wilcox, C. R., Dageforde, M. L. and Jirak, G. A., "MAINSAIL Language Manual," supported by Nat'l Inst's of Health Grant RR-00785, (Stanford, CA: S.U. Medical Center Computer Facility, July, 1979).

"An Introduction to PADL," Memo TM-22, (Rochester, NY: U. Rochester, Production Automation Proj., Dec. 1974).

"User's Guide to VAL," (Danbury, CT: Unimation Inc., Feb. 1979).

Wang, D., "Some Comments and Information on PEARL," MAT Memo 6031, May 24, 1978.

"Preliminary ADA Reference Manual," ACM SIGPLAN Notices, Vol. 14, No. 6, Part A (June 1979).

"Rational for the Design of the ADA Programming Language," ACM SIGPLAN Notices, Vol. 14, No. 6, Part B (June 1979).

Safety

Park, W. T., "Robot Safety Suggestions," Tech. Note 159 (Menlo Park, CA: SRI Int'l, Artif. Intel. Ctr., April 1978).

"Detecting Servo Failures with Software," Tech. Supp. Package FRC-11003 (Wash. D.C.: Nat'l Aeron. & Space Admin., Technol. Util. Branch), summary in NASA Tech Briefs (Fall 1978), p. 415.

Sensor Technology and Applications

Burton, R. P., "Real-Time Measurement of Multiple three-Dimensional Positions," Ph. D. Thesis, Order No. 73-21,248 (Salt Lake City, UT: U. Utah, Comp. Sci. Dept., 1973), 129 pp.

Coles, L., "Decision Analysis for an Experimental Robot with Unreliable Sensors," 4th Int'l Jt. Conf. on Artificial Intelligence, Tbilisi, Georgia, U.S.S.R., 3-8 Sep. 1975 (Cambridge, MA: The Artif. Intell. Lab., Publications Dept., 545 Technol. Sq.), pp. 749-757.

Durham, T., "Infrared Light for a New Wireless Revolution," New Scientist, Vol. 84 (double issue) No. 1186/1187, Dec. 20/27, 1979, pp. 931-933.

Koskinen, K., and Niemi, A., "Object Recognition and Handling in an Industrial Robot System," Proc. 8th Int'l Symp. on Indust. Robots, Vol. II, Stuttgart, 30 May - 1 June, 1978 (Bedford, England: Int'l. Fluidics Services, Ltd.), pp. 744-755.

Lewis, R. A. and Johnson, A. R., "A Scanning Laser Rangefinder for a Robotic Vehicle," 5th Int'l Jt. Conf. on Artificial Intelligence, Cambridge, MA, 22-25 Aug., 1977 (Pittsburgh, PA: Carnegie-Mellon U., Dept. of Comp. Sci.), pp. 762-768.

Nitzan, D., "The Measurement and Use of Registered Reflectance and Range Data in Scene Analysis," Tech. Note 128, Nat'l Sci. Found. Grant ENG-75-09327, Adv. Res. Projects Agency Contract DAHC04-72-C-0008 (Menlo Park, CA: SRI Int'l, Artif. Intel. Ctr., April 1976).

Sato, N., Heginbotham, W. B., and Pugh, A., "A Method for Three Dimensional Part Identification by Tactile Transducer," Proc. 7th Int'l Symp. on Indust. Robots, Tokyo, 19-21 October, 1977 (Tokyo: Japan Indust. Robot Assoc.), pp. 577-585.

Watson, P. and Drake, S., "Pedestal and Wrist Force Sensors for Automatic Assembly," Report P-176 (Cambridge, MA: C. S. Draper

Labs, Inc., June 1975), and in Proc. 5th Int'l Symp. on Indust. Robots, Chicago, IL, 22-24 September, 1975 (Chicago, IL: IIT Research Inst.), pp. 501-512.

Yasaki, E. K., "Voice Recognition Comes of Age," Datamation, August 1976, pp. 65-68.

"Connected Speech Recognition System," product brochure, Nippon Electric Co. (Melville, NY: NEC America, Inc., 1979).

"Electrically-Controlled Variable-Color Optical Filters," Tech. Supp. Package MSC-14944 (Wash. D.C.: Nat'l Aeron. & Space Admin., Technol. Util. Branch), summary in NASA Tech Briefs (Spring 1977), pp. 59-60.

"Non-contacting Ultrasonic Distance Measurement," (U.K. Patent Application 803/75) NRDC Inventions (London, England: Nat'l Res. & Devel. Corp., July 1976).

Wang, S. S. and Will, P. M., "Sensors for Computer Controlled Mechanical Assembly," in The Industrial Robot.

Standardization Issues in Robotics

Connolly, R., "PDP-11 Chosen as Tactical Base," Electronics, 14 Oct. 1976, p. 77.

Evans, J. m., Barbera, A. J., and Albus, J. S., "Standards and Control Technology for Industrial Robots," Proc. 7th Int'l Symp. on Indust. Robots, Tokyo, 19-21 October, 1977 (Tokyo: Japan Indust. Robot Assoc.), pp. 479-486.

Evans, J., "CAM Standards Directions," Paper MS78-483 (Dearborn, MI: Soc. Manuf. Eng.'s, Mkt. Svc. Dept., 1978), 10 pp.

Hasegawa, K., and Kaneko, T., "Study on the Standardization of Terms and Symbols Relating to Industrial Robots in Japan," Proc. of 3rd CISM-IFTOMM Symp. on Theory and Practice of Robots and Manipulators, Udine, Italy, 12-15 Sep. 1978, (New York: Springer-Verlag, 1978), pp. 471-478.

Yoosufani, Z., and Boothroyd, G., "Design for Manufacturability, Report No. 2, Design of Parts for Ease of Handling," Master's Thesis, Nat'l Sci. Found. Grant APR-77-10197 (Amherst, MA: U. Mass., Dept. Mech. Eng., Sep. 1978).

"Standards for Computer-Aided Manufacturing, Final Technical Report, March through December 1977," (Dayton, OH: Manuf. Technol. Div., Air Force Materials Lab., Wright-Patterson AFB, 1977) and (Wash., D.C.: Natl. Bur. Stds., 1977).

"Study on Standardization of Industrial Robots," (Tokyo: Japan Indust. Robot Assoc., Jan. 1976).

"Workshop on Standards for Image Pattern Recognition, 3-4 June, 1976", Publ. 500-8 (Wash., D.C.: Natl. Bur. Stds., June 1976).

Surveys of Artificial Intelligence

Barr, A., and Feigenbaum, E. A., Handbook of Artificial Intelligence (Stanford, CA: Stanford U. Comp. Sci. Dept., 1980).

Boden, M. A., Artificial Intelligence and Natural Man (New York: Basic Books, 1977).

Charniak, E., Riesbeck, C., and McDermott, D. Artificial Intelligence Programming (Hillsdale, NJ: Lawrence Erlbaum Associates., 1979).

Jackson, P. C., Jr., Introduction to Artificial Intelligence (New York: Petrocelli Books, 1974).

McCorduck, P., Machines Who Think (San Francisco: W. H. Freeman, 1979).

Minsky, M. (ed.), Semantic Information Processing (Cambridge, MA: The MIT Press, 1968).

Nilsson, N. J., Problem Soving Methiods in Artificial Intelligence (New York, NY: McGraw-Hill, 1971).

Nilsson, N. J., Principles of Artificial Intelligence (Palo Alto, CA: Tioga Publishing Co., 1980).

Raphael, B., The Thinking Computer: Mind Inside Matter (San Francisco: W. H. Freeman, 1976).

Shapiro, S., Techniques of Artificial Intelligence (New York: D. Van Nostrand, 1979).

Winograd, T., "Five Lectures on Artificial Intelligence," presented at Electrotechincal Lab., Tokyo, Japan, 18-23 March 1974, supported by the Electrotechnical Lab., and Adv'd Res. Proj's Agency Contract DAHC15-73-C-0435, Memo AIM-246, and Report STAN-CS-74-459 (Stanford, CA: Stanford Artif. Intel. Proj., Sep. 1974).

Winston, P. H., Artificial Intelligence (Reading, MA: Addison-Wesley, 1977).

Winston, P. H., and Brown, R. H. (eds.), Artificial Intelligence: an MIT Perspective, 2 vols. (Cambridge, MA: MIT Press., 1979).

Teleoperator Systems and Techniques

Brooks, T. L., "SUPERMAN: A System for Supervisory Manipulation and the Study of Human/Computer Interactions," Master's Thesis Grant 04-7-158-440079, M.I.T. Sea Grant Program, Office of Naval Res. contract No. N00014-77-C-0256 (Cambridge, MA: M.I.T. Mech. Eng. Dept., May 1979).

Goome, R., Jr., "Force Feedback Steering of a Teleoperator System," Report T-575 (Cambridge, MA: C. S. Draper Labs, Inc., Aug. 1972).

Sheridan, T., "Supervisory Control of Remote Manipulators for Undersea Applications," _Proc. Int'l Conf. on Cyb. & Soc._, Wash., D. C., 19-21 Sep., 1977 (Piscataway, NJ: IEEE Svc. Ctr.), pp. 237-242.

Katowski, J., "Remote Manipulators as a Computer Input Device," (Chapel Hill, NC: U. N. C., Dept. Comp. Sci., 1971).

Miscellaneous References

"Technical Manual, VERSATRAN Automation System, Model 600 Control Unit," (Herndon, Va.: AMF Electrical Products Development Division, Nov., 1977).

"Introducing the STC 4305 Solid State Disk," product brochure (Louisville, CO: Storage Technology Corporation, 1979).

"Cartesian 5, the One-Station System for Automated Plastic Trimming," product brochure on a trainable Cartesian-geometry tool carrier (Dale, IN: Thermwood Machinery Co., Inc., 1978).

Updates on Technology, Vol. 1, No. 5 (Dale, IN: Thermwood Machinery Co., Inc., Oct./Nov. 1978).

Part II

Robotics in Motor Vehicle Manufacture

The information in Part II is from *Robotics Use in Motor Vehicle Manufacture,* prepared by W.R. Tanner of Productivity Systems Inc. and W.F. Adolfson of Advanced Technology, Inc. for the U.S. Department of Transportation, February 1982.

-1-
Introduction

The automobile industry has, historically, been a major developer and user of automation and industrial robots. In 1909, Henry Ford automated the movement of automobiles through assembly operations by adapting the continuously moving conveyor (which he had seen used in the meat packing industry) for transporting the chassis of Model T automobiles past a series of work stations. In 1922, A.O.Smith developed a mechanized automobile frame manufacturing plant.

In 1925, F.G.Wollard developed, in England, the concept of the transfer machine, that is, automatically transferring parts through a series of machine tools. The concept was first used in 1927 by Morris Motors of England; it was technically but not economically successful. In the 1950's, George C. Devol developed the concept of the digitally controlled manipulator, or industrial robot. A patent was granted in 1961 to Devol for "Programmed Article Transfer" and in the same year, the first production installation of an industrial robot was made in a U.S. automobile manufacturer's die casting plant.

The years since the 1900's, the 1920's and even since the 1960's have seen significant technological changes in automation and robots and a steady increase in their utilization in the automobile industry worldwide. The 1980's will see continued technological changes and a continued increase in utilization, especially in industrial robots.

For instance, transfer lines have borrowed the technology of the robots; the "programmable transfer line" is now a reality. The robot population in the U.S. automobile industry, which was about 1,065 units at year end 1980, may exceed 35,000 units by 1990 (General Motors alone estimates that its robot population may reach 14,000 by 1990). New developments in sensory control and other specialized capabilities will enable robots to perform many tasks in the automobile factories which presently must be done by humans. Microelectronics and computers will facilitate automating of other manual operations.

The improved capabilities of robots and automation, in combination with an increase in their utilization, will reduce the cost of manufacturing automobiles. The number of manhours of direct labor per car will be reduced, while the number of manhours per car of indirect labor to maintain and repair the automation and robots will increase (but not in the same proportion as the direct labor reduction). Capital investment requirements for automation and robots, especially through the mid-1980's, will be high.

-2-

Industrial Robot Technology

The Robot Institute of America (RIA), which is the U.S. trade association representing the manufacturers, distributors, suppliers and major users, defines an industrial robot as "a reprogrammable multi-functional manipulator designed to move material, parts, tools or specialized devices, through variable programmed motions for the performance of a variety of tasks".

The RIA further classifies industrial robots by type, as follows:

TYPE A: Programmable, Servo-Controlled, Point-to-Point

TYPE B: Programmable, Servo-Controlled, Continuous-Path

TYPE C: Programmable, Non-Servo Robots for General Purpose

TYPE D: Programmable, Non-Servo Robots for Die Casting and Molding Machines

TYPE E: Non-Programmable, Mechanical Transfer Devices (Pick-and-Place)

Industrial robots, regardless of type, generally consist of several major elements: the manipulator or "mechanical unit" which actually performs the manipulative functions; the controller or "brain" which stores data and directs the movement of the manipulator and the power supply which provides energy to the manipulator.

The manipulator is a series of mechanical linkages and joints capable of movement in various directions to perform the work of the robot. These mechanisms are driven by actuators which may be pneumatic or hydraulic rotary or linear actuators or electric motors. The actuators may be coupled directly to the mechanical links or joints or may drive indirectly through gears, chains or ball screws. In the case of pneumatic or hydraulic drives, the flow of air or oil to the actuators is controlled by valves mounted on the manipulator.

Feedback devices are installed to sense the positions of the various links and joints and transmit this information to the controller. These feedback devices may simply be limit switches actuated by the robot's arm or position measuring devices such as encoders, potentiometers or resolvers and/or tachometers to measure speed. Depending on the devices used, the feedback data is either digital or analog.

The controller has a three-fold function: first, to initiate and terminate motions of the manipulator in a desired sequence and at desired positions; second, to store position and sequence data in memory; and third, to interface with the "outside world".

Robot controllers run the gamut from simple step sequencers through pneumatic logic systems, diode matrix boards, electronic sequencers to microprocessors and minicomputers. The controller may be either an integral part of the manipulator or housed in a separate cabinet.

The complexity of the controller both determines and is determined by the capabilities of the robot. Simple non-servo (Types C and D) devices usually employ some form of step sequencer. Servo-controller (Types A and B) robots use a combination of sequencer and data storage (memory). This may be as simple as an electronic counter, patch board or diode matrix and series of potentiometers or as sophisticated as a minicomputer with core memory. Other memory devices employed include magnetic tape, magnetic disc, plated wire and semiconductor (solid state RAM). Processor or computer based controller operating systems may be hard wired, stored in core memory or programmed in ROM (read only memory).

The controller initiates and terminates the motions of the manipulator through interfaces with the manipulator's control valves and feedback devices and may also perform complex arithmetic functions to control path, speed and position. Another interface with the outside world provides two-way communications between the controller and ancillary devices. This interface allows the manipulator to interact with whatever other equipment is associated with the robot's task.

The function of the power supply is to provide energy to the manipulator's actuators. In the case of electrically driven robots, the power supply functions basically to regulate and filter the incoming electrical energy. Power for pneumatically actuated robots is usually supplied by a remote compressor, which may also service other equipment.

Hydraulically actuated robots normally include a hydraulic power supply as either an integral part of the manipulator or as a separate unit. The hydraulic system generally follows straightforward industrial practice and consists of an electric motor driven pump, filter, reservoir and, usually, a heat exchanger (either air or water).

There are significant differences in operation, capability and, to some extent, physical characteristics among the various types of industrial robots. The typical operating sequence of a Type A or Type B robot (programmable, servo-controlled, point-to-point or continuous-path) is as follows:

> Upon start of program execution, the controller addresses the memory location of the first command position and also reads the actual position of the various axes as measured by the position feedback system.
>
> These two sets of data are compared and their differences, commonly called "error signals", are amplified and transmitted as "command signals" to servo valves for the actuator of each axis.

The servo valves, operating at constant pressure, control flow to the manipulator's actuators; the flow being proportional to the electrical current level of the command signals.

As the actuators move the manipulator's axes, feedback devices such as encoders, potentiometers, resolvers and tachometers send position (and, in some cases, velocity) data back to the controller. These "feedback signals" are compared with the desired position data and new error signals are generated, amplified and sent as command signals to the servo valves.

This process continues until the error signals are effectively reduced to zero, whereupon the servo valves reach null, flow to the actuators is blocked and the axes come to rest at the desired position.

The controller then addresses the next memory location and responds appropriately to the data stored there. This may be another positioning sequence for the manipulator or a signal to an external device.

The process is repeated sequentially until the entire set of data, or "program" has been executed.

The significant features of Type A and B robots are:

The manipulator's various members can be commanded to move and stop anywhere within their limits of travel, rather than only at the extremes.

Since the servo valves modulate flow, it is feasible to control the velocity, acceleration and deceleration of the various axes as they move between programmed points.

Generally, the memory capacity is large enough to store many more positions than a non-servo robot.

Both continuous path and point-to-point capabilities are possible.

Accuracy can be varied, if desired, by changing the magnitude of the error signal which is considered zero. This can be useful in "rounding the corners" of high-speed continuous motions.

Drives are usually hydraulic or electric and use state-of-the-art servo control technology.

Programming is accomplished by manually initiating signals to the servo valves to move the various axes into a desired position and then recording the output of the feedback devices into the memory of the controller. This process is repeated for the entire sequence of desired postions in space.

Common characteristics of Type A and B robots include:

Smooth motions are executed, with control of speed and, in some cases, acceleration and deceleration. This permits the controlled movement of heavy loads.

Maximum flexibility is provided by the ability to program the axes of the manipulator to any position within the limits of their travel.

Most controllers and memory systems permit the storage and execution of more than one program, with random selection of programs from memory via externally generated signals.

With microprocessor or minicomputer based controllers, subroutining and branching capabilities may be available. These capabilities permit the robot to take alternative actions within the program, when commanded.

End-of-arm positioning accuracy of 1.5 mm (.060 in.) and repeatability of ± 1.5 mm (± .060 in.) are generally achieved. Accuracy and repeatability are functions of not only the mechanisms, but also the resolution of the feedback devices, servo valve characteristics, controller accuracy, etc.

Due to their complexity, Type A and B robots are more expensive and more involved to maintain than Type C and D robots and tend to be somewhat less reliable.

Type A, programmable, servo-controlled, point-to-point robots are used in a wide variety of industrial applications for both parts handling and tool handling tasks.

Significant features of the Type A robot are:

For those robots employing the "record-playback" method of teaching and operation, initial programming is relatively fast and easy; however, modification of programmed positions cannot be readily accomplished during program execution.

Those robots employing sequencer/potentiometer controls tend to be more tedious to program; however, programmed positions can be modified easily during program execution by adjustment of potentiometers.

The path through which the various members of the manipulator move when traveling from point-to-point is not programmed or directly controlled in some cases and may be different from the path followed during teaching.

Common characteristics of Type A robots include:

High capability control systems with random access to multiple programs, subroutines, branches, etc., provide great flexibility to the user.

These robots tend to lie at the upper end of the scale in terms of load capacity and working range.

Hydraulic drives are most common, although some robots are available with electric drives.

The second type of servo-controlled robot is the Type B, programmable, servo-controlled, continuous-path. Typically, the positioning and feedback principles are as described previously, that is, common with the Type A robots. There are, however, some major differences in control systems and some unique physical features.

The significant features of the Type B robot are:

During programming and playback, data is sampled on a time base, rather than as discretely determined points in space. The sampling frequency is typically in the range of 60 to 80 Hz.

Due to the high rate of sampling of position data, many spatial positions must be stored in memory. A mass storage system, such as magnetic tape or magnetic disc is generally employed.

During playback, due to the hysteresis of the servo valves and inertia of the manipulator, there is no detectable change in speed from point-to-point. The result is a smooth continuous motion over a controlled path.

Depending upon the controller and data storage system employed, more than one program may be stored in memory and randomly accessed.

The usual programming method involves physically moving the end of the manipulator's arm through the desired path, with position data automatically sampled and recorded.

Speed of the manipulator during program execution can be varied from the speed at which it was moved during programming by playing back the data at a different rate than that used when recording.

Type B robots share the following characteristics:

These robots generally are of smaller size and lighter weight than point-to-point robots.

Higher end-of-arm speeds are possible than with point-to-point robots; however, load capacities are usually less than 10 kg (22 lbs.).

Their common applications are to spray painting and similar spraying operations, polishing, grinding and arc welding.

The programmable, non-servo robots, Type C and D also share some common characteristics between themselves and operate in the same basic manner. A typical operating sequence of a Type C or D hydraulic or pneumatic actuated robot is as follows:

Upon start of program execution, the sequencer/controller initiates signals to control valves on the manipulator's actuators.

The valves open, admitting air or oil to the actuators and the members begin to move.

The valves remain open and the members continue to move until physically restrained by contact with end stops.

Limit switches signal the end of travel to the controller which then commands the control valves to close.

The sequencer then indexes to the next step and the controller again outputs signals. These may again be to the control valves on the actuators or to an external device such as a gripper.

The process is repeated until the entire sequence of steps has been executed.

The significant features of a Type C or D robot are:

The manipulator's various members move until the limits of travel (end stops) are reached. Thus there are usually only two positions for each axis to assume.

The sequencer provides the capability for many motions in a program, but only to the end points of each axis.

Deceleration at the approach to the stops may be provided by valving or shock absorbers.

It is feasible to activate intermediate stops on some axes to provide more than two positions; however, there is a practical limit to the number of such stops which can be installed.

The programmed sequence can be conditionally modified through appropriate external sensors; however, this class of robots is restricted to the performance of single programs.

Programming is done by setting up the desired sequence of moves and by adjusting the end stops for each axis.

Common characteristics of non-servo robots include:

Relatively high speed may be possible, due to the generally smaller size of the manipulator and full flow of air or oil through the control valves.

Repeatability to within 0.25 mm (0.010 in.) is attainable on the larger units and to within 0.1 mm (0.004 in.) or better on smaller units.

These robots are relatively low in cost; simple to operate, program and maintain; and are highly reliable.

These robots have limited flexibility in terms of program capacity and positioning capability.

The differences between Type C and Type D robots are minimal, although the two types are not necessarily interchangeable in all applications. The Type D robots are generally constructed to mount directly onto a die casting or plastic molding machine and provide only as many axes of motion as are necessary to perform a single function, namely, to extract a part from the machine and drop it into a container or onto a conveyor. In many cases, a Type C robot could perform the same function as a Type D but might have more capability than was required for the task and might not, therefore, be as cost effective or as efficient.

There are hundreds of manufacturers of robots worldwide and their products provide a wide range of capabilities relative to performance, that is, load capacity, speed, positioning accuracy (repeatability), flexibility, etc. There is no "ideal" robot, capable of performing all tasks equally well. The user must, therefore, establish parameters of performance and choose the robot which best meets these requirements.

There are definite tradeoffs in performance capabilities, as with most mechanical devices. For example, robots capable of handling heavy loads usually do not move rapidly. In fact, load capacity specifications for robots often quote nominal full-speed payload capacity and maximum payload capacity, at reduced speed. The maximum payload capacity may be as much as 50% greater than nominal, at 30% to 50% lower speed.

Another area where performance tradeoffs are made is in positioning repeatability, which is often referred to, erroneously, as "accuracy". There are a number of factors which affect repeatability, including speed, load, drive system and positioning system. Some of these factors, in turn, trade off with other performance parameters.

The best positioning repeatability is attained by using physical stops to establish axes positions. This is the positioning system used on non-servo (Types C and D) robots. Such robots, however, lack the multi-position flexibility of the servo-controlled (Types A and B) robots.

The following table summarizes the significant trade offs:

TABLE 1. - TRADE OFFS IN ROBOT PERFORMANCE CAPABILITIES

Capability	Significant Trade Offs
Speed	Repeatability, load capacity, Size of work envelope
Repeatability	Speed, load capacity, Size of work envelope, Flexibility
Load Capacity	Repeatability, Speed
Size of Work Envelope	Repeatability, Speed
Flexibility	Low Cost, Simplicity
Reliability	Flexibility

An issue related to trade offs is that of the relative merits of general purpose and special purpose robots. Here again, the basic difficulty is that no "perfect" robot exists -- there is no robot which can perform all automobile manufacturing tasks well -- and trade offs must be made for many applications. Presently, the trend of the robotics industry is to develop special purpose robots for applications with a sufficiently large market (or potential market).

For the user, the availability of special purpose robots can be of significant advantage. Applications engineering is usually simpler and less costly than it would be if a general purpose robot were used. The reliability of the operation is generally higher because the major element of the system is a more-or-less "standard" robot. The efficiency of the operation is generally higher because the special purpose robot's capabilities are matched to the requirements of the task.

However, the user is giving up some flexibility; although the robot can be re-applied on a similar operation if the original task is discontinued, it cannot be used on a significantly different job. That is, a painting robot, for example, can be used to spray other materials, such as undercoating, but cannot be used for spot welding or material handling.

Presently, a number of examples of such special purpose robots exist in the automobile industry. These include robots for spraying, arc welding and small-component assembly. The following table summarizes the significant characteristics of each of these special purpose robots. For comparison, the significant characteristics of a typical general purpose robot used by the automobile industry are also shown:

TABLE 2. - UNIQUE CHARACTERISTICS OF SPECIAL PURPOSE ROBOTS IN AUTOMOBILE MANUFACTURING

Robot Purpose	Unique Characteristics
Spraying	Direct lead-through programming
	Continuous-path, trajectory control playback
	Light Payload (15 lbs.)
	Low inertia, high speed arm (60 in. per second or more)
	Large memory (up to 20 minutes)
	Intrinsically safe in explosive atmosphere
Arc Welding	Point-to-point programming
	Continuous-path, trajectory control playback
	Precise, low speed traverse rates (± 1% of programmed speed)
	Linear or circular interpolation
	Good positioning repeatability (± .010 in.)
	Up to seven axes of motion/control
Assembly	Anthropomorphic size and configuration
	High speed motions (50-60 inches per second)
	Good positioning repeatability (± .005 inches)
	Computer control
	Sensory feedback capability (vision, force, touch, etc.)
	Off-line programming, high-level programming language
General	Point-to-point programming, point-to-point playback
	Heavy payload (up to 250 lbs.)
	Any configuration (cartesian, cylindrical, polar or anthropomorphic)
	Limited memory (about 1000 points)
	Moderate repeatability (± .050 inch)
	Electronic control

There are a number of developments in industrial robots which are likely to evolve in the near future. Some of these will come from user development programs, some from robot manufacturers and some from independent research organizations. Regardless of the source, these developments will come in response to present or future requirements, which will be identified and described by the potential users, including the automobile industry.

One area of continued development will be an increasing number of special-purpose robots. These will include programmable devices for loading and unloading stamping presses, small high-speed assembly robots and medium-sized, electric drive assembly robots.

Devices presently used in the U.S. for loading and unloading stamping presses and for transferring stampings from press to press include point-to-point, servo-controlled (Type A) robots, point-to-point, non-servo (Type C) robots and non-programmable mechanical transfer (Type E) devices. The Type A robots are generally too costly for these applications and have marginal speed and positioning repeatability. Except for very small units capable of handling less than a one pound payload, the Type C robots are too slow for these applications. The Type E devices lack the manipulative capability required for some loading operations and also lack the flexibility to accommodate rapid changes.

Some relatively fast, simple Type C robots have been developed in Japan by Aida, Daido and Toshiba Seiki, primarily for press work. The speed and precision of these devices are impressive: handling of stampings up to four by eight feet in size, weighing 35 to 40 pounds on multiple-press lines at cycles as short as five seconds for transfer and press actuation is not uncommon. Comparable operations in U.S. stamping facilities are accomplished only with hard automation on dedicated (single-product) press lines. These Japanese press-tending robots are generally pneumatically actuated and are priced in the low to mid $20,000 range. At this time, none of these robots are being marketed in the U.S. and only one is being distributed in Europe.

Another special purpose robot which would be applicable for automotive operations is a high-speed assembly robot. There are numerous small, servo-controlled, electric drive (Type A) "assembly" robots available in the U.S., Europe and Asia. These units have adequate positioning repeatability and most, being controlled by a microcomputer, have a high degree of flexibility, making them particularly adaptable for batch manufacturing. There are also numerous small, non-servo, high-speed (Type C) "pick-and-place" robots, useful for high volume production, available. These robots are so lacking in flexibility and capability, however, that it is common practice to dedicate each robot to the handling of a single part or the performance of a single step in an assembly operation.

For the relatively high production volumes common in the assembly of small components for automobiles, what is required is a robot capable of accurate movement and positioning at speeds two to three times as fast as the 20 to 30 inches per second characteristic of present Type A "assembly" robots. This new assembly robot should also be capable of executing several routines as part of its normal task and be easily programmed to compensate for minor changes in its operational environment or to perform new tasks.

The development of such a robot is within the scope of the current state-of-the-art and at least two approaches would be feasible. One approach would be the improvement of the power and stiffness of drives and mechanisms to produce high-speed movements with good positioning repeatability. The power and precision required in the mechanism would probably result in a significant increase in price compared to present

Type A "assembly" robots. Another approach would involve reduction of the mass of the mechanism to increase speed and automatic compensation for positioning inaccuracies likely to result from a lack of mechanical rigidity or repeatability. This would require new software, increased control capability, initialization or "zeroing" routines and, perhaps, higher resolution position feedback devices.

A third special purpose robot is a medium sized, servo-controlled (Type A) robot for assembly. The "assembly" robots presently available are mostly relatively small devices having typically, a spherical working volume of less than 36 inches radius and a payload capacity (part plus end-effector) of less than 10 pounds. A few slightly larger robots, including the ASEA IRb6 and Hitachi Process Robot are available, but these are significantly more costly than the smaller units (up to $60,000 or more) and also are marginal in payload capacity (about 20 pounds maximum).

There are significant numbers of, primarily, mechanical assembly operations, which require the larger work envelope and higher payload capacity of this new medium sized assembly robot. These include the assembly of heads, blocks and engines, transmissions, differential units, suspension systems, steering gear and columns, brakes, air conditioning compressors, etc. The new medium sized robot capable of performing these tasks should be electrically driven and servo-controlled (Type A), with a payload capacity of 25 pounds to 50 pounds and a work envelope having a radius of at least 42 inches. It should be micro-computer controlled, with a good positioning repeatability (± .005 inches) and should move at speeds up to 50 inches per second. The cost should be around $50,000.

The development of the foregoing new robots will, at least in the U.S., come from manufacturers and distributors of robots. The prospective users (that is, the automobile industry) may, however, have significant input in the form of performance specifications and perhaps shared funding for new product development.

Another likely area of robotic development will be in control and monitoring, particularly hierarchical systems involving numbers of robots performing a common task. A current example of this is General Motors' NC Painter, which uses a hierarchy of computers to control as many as 10 robots, allocating programs to each, monitoring performance and reallocating programs whenever a robot is down for any reason. Other robotics applications where centralized control and monitoring are applicable include spot welding, transfer machine loading and unloading and assembly.

In spot welding operations, it is common practice to install a number of robots, perhaps as many as twenty, along with a transfer and positioning system to move and orient the sheet metal assemblies being welded. At present, if any of the robots is unable to perform its welding tasks, due to malfunction of it or of its welding equipment, the missed welds are made either manually or with a "standby" robot at the end of the transfer system. Activation of the "standby" robot and selection of the welding program it is to execute is generally done manually.

With a centralized control and monitoring computer, the activation of the "standby" robot could be automatic. Alternatively, a more effective approach would be for the central computer, upon detection of a robot or welding equipment malfunction, to redistribute that robot's task among the remaining robots on the line, thereby eliminating the need for "standby" robots. In most cases, this could be accomplished without delay and without significant impact upon the normal cycle time of the system. The monitor could also shut down a robot for periodic maintenance (particularly of the welding gun) to assure quality welds and prevent breakdowns, according to a predetermined schedule or whenever there were indications of declining weld quality.

Similarly, in transfer machine loading and unloading, the centralized computer could direct the robots under its control to transfer parts to or from storage buffers or between parallel machine lines. In this way, it would maintain relatively constant output despite interruptions on various segments of the system.

In assembly operations, the central computer could monitor and direct the robots under its control in a manner similar to the transfer machine systems. In addition, for batch assembly operations, the centralized control could down-load new programs for the robots whenever there was a change in the product being assembled.

The basic programs and interconnections for such hierarchical computer systems for robot monitoring and control are and will continue to be formulated by basic research and development organizations. Much of the necessary hardware and software already exists or has been described. However, implementation of these techniques will be accomplished by the user or by the supplier of the total system. It is unlikely that the robot manufacturers and distributors, unless they are also in the systems business, will contribute significantly to these developments.

Another development area, one where there is presently a great deal of activity, is that of sensory feedback for robots. The industrial robots currently in use in automobile manufacturing and assembly operations are "deaf, dumb and blind" and perform their task entirely by rote. Thus, a repeatable, orderly environment is necessary, one which requires mechanisms to transfer, orient and position the work for the robot. This is often in complete contrast to the work handling methods which are provided for manual performance of the same tasks.

Sensory systems under development include vision, tactile and force sensing, with the systems including the sensors, electronic interfaces with the robot controls and robot controller software. Much additional development is needed on all three elements of the systems.

Of the three most common types of robot sensory feedback systems, vision, tactile and force, the vision capability seems to offer the greatest potential flexibility for robot applications. In the automotive field, robot vision could be used effectively for a wide variety of applications, including spot welding, arc welding, assembly, machine tending, inspection and general material handling. Robot vision is, in fact, considered a key element in the successful implementation of most arc welding and assembly operations and in some machine tending and material handling operations as well. Vision systems are also the most complex to develop and implement.

Present vision systems require special lighting and/or movement of parts, special electronic interfaces between camera and data processor and relatively long data processing time. They lack depth or range perception or good image resolution. The vision systems which are commerically available or being developed are very robot-specific; that is, they work with only certain models of robots. Image data processing time, the time required for the system's electronics to extract useful information from the "picture" the vision system sees and transmit appropriate data and commands to the robot controller, is generally so long that real-time control of the robot's movements are not feasible.

Further development is required in all elements of robot vision systems. Resolution, light intensity limits, gray scale and color discrimination capabilities of solid-state imaging devices (cameras) need improvement. New and modified data extraction, data processing and image reconstruction algorithms and methods must be devised to increase the speed of the systems. Real-time, high-speed interfaces with the robot controls need to be developed to permit control of a robot's movements with external feedback devices at normal operating speeds. System hardware and software must be made less equipment-specific so that any vision system can work with any robot.

Force feedback is another sensory system requiring significant development. As with vision, force feedback systems today exist primarily in laboratories and research facilities. Such systems will be useful in assembly, machine tending, foundry, deburring and polishing operations.

Developments required include rugged, simple force sensors with good sensitivity over a wide force range and real-time, high-speed interfaces with robot controls. Force feedback systems should be capable of being easily programmed to control the magnitude and direction of a force exerted by the robot arm, under varying conditions.

Tactile sensors may be somewhat simpler to develop and implement; some rudimentary tactile sensing is already being used for removing parts or material from racks or stacks. In this area of sensory feedback, development is now centered upon sensor arrays capable of extracting shape information from contact with a part. Such capability could enhance robot applications to machine tending, assembly and material handling operations through determination of part position and orientation while the part was being handled. As with other sensory feedback systems, further development of sensors is required, as well as real-time, high-speed interfaces with robot controls. In the U.S., the development of sensory feedback systems is presently concentrated primarily in basic research organizations, with limited activity also in user research facilities and among some robot, electronics and sensor manufacturers. Because all of these sensory feedback systems require the integration of several technologies and combinations of various commercial and special components, it is not likely that either users or robot manufacturers will take over the basic research and development or the commercialization of such systems. In the near future, it is probable that sensor manufacturers will develop and market "plug-in" sensory feedback systems for robots, using the results of the continued development efforts of the basic research organizations.

The situation in Europe and Japan is somewhat different. There, fundamental developments in sensory feedback technology are undertaken in basic research organizations, but the application of these fundamentals, that is, the development of sensory systems, is primarily handled by the manufacturers of robots (who, incidently may also be the major users of these robots). Thus, the integration of robots and sensory feedback may evolve more quickly. On the other hand, the development of more universal, "plug-in" sensory feedback systems is less likely under this condition.

An important area of future development will be the integration of robotics into computer-aided design/computer-aided manufacturing (CAD/CAM) systems. One of the first steps in this integration process will be the development of off-line programming for robots, that is, the preparation of sets of instructions for robots at a computer terminal, using a vocabulary of commands. Off-line programming vocabularies or "languages" have already been developed by the robot manufacturers or by basic research organizations for a few specific robots. However, each language is different from the others and none are compatible with existing CAD/CAM system languages.

The integration of robotics with CAD/CAM, therefore, will require the development of translation programs, which will be equipment-specific (much like the post-processor programs for NC machines). These programs will take a set of general instructions for a robot task, developed with a CAD or CAM computer, and convert them into specific commands for a specific robot in a particular location in the factory. The advantage of this approach is that robot programmers will not have to learn a whole series of languages, one for each kind of robot.

There are several significant benefits to be gained from off-line programming or robots. In robotic painting applications for example, the robots are currently programmed by guiding the robot through the desired motions, on location in the paint booth. This requires a skilled spray painter who has also been instructed in robot programming. It also necessitates either interrupting production or programming on weekends or during the night after production has been stopped. Programming of other robot tasks such as machine tending, spot welding, assembly, etc. also must be done on location. Where the robot's tasks are complex, as in spot welding or assembly, the programming on-site can take a great deal of time. Simulation or tryout cannot be accomplished without actually installing the robots in the work place or constructing a detailed mockup of the work place. Off-line programming and integration with CAD/CAM systems will largely overcome these problems.

The development of off-line programming capabilities will be undertaken primarily by the manufacturers of robots. This will not, however, resolve the problem of equipment-specific languages or facilitate the integration of robotics with CAD/CAM. The development of translation and simulation programs and integration with CAD/CAM will thus be done primarily by manufacturers of CAD/CAM computers and systems, by major robot users, or by third party systems contractors.

Another important development area is the establishment of "standards" for industrial robots. Such standards will include mechanical interfaces for mounting end-of-arm devices on the robots, electrical interfaces for interconnections between the robot controller and the external devices with which the robot interacts, sensor interfaces for sensory feedback systems and computer interfaces for hierarchical control systems, monitoring and off-line programming. Other standards will include the programming languages themselves, performance evaluation, testing and reporting of results and terminology used to describe the robots, their configurations, their components and their performance.

The National Bureau of Standards has already sponsored a work shop on robotic standards and is funding the development of a glossary of robot terminology. It is likely that NBS will continue in its leadership role in this development area, although actual standards will probably be developed, reviewed and published by committees made up of robot users, manufacturers and NBS representatives.

In Europe and Japan, some standardization development is already underway. A government supported research organization in West Germany is involved in the systematic testing and evaluation of robots, according to standard procedures which they developed. Results of tests on a number of robots have already been published. The Japan Industrial Robot Association has developed and published a glossary of terms and is developing evaluation and testing standards and performance specifications standards.

The following table summarizes some of the current and potential application areas for robots in automobile manufacturing and the related developments which are either necessary or highly desirable to enhance these applications.

TABLE 3. - FUTURE DEVELOPMENTS IN ROBOTS AND ASSOCIATED EQUIPMENT TO ENHANCE AUTOMOBILE MANUFACTURING OPERATIONS

Application	Development
Metal Forming	Special purpose robot - high speed Type C, medium to large size, low cost
Spot Welding	Sensory feedback system - simple vision
	Computer control and monitoring for multiple-robot systems
	Computer aided design - off-line programming
	Standardized mechanical, electrical and computer interfaces
Arc Welding	Sensory feedback system - vision based joint tracking
Component Manufacturing	Sensory feedback systems - simple vision, force feedback, tactile
	Computer aided design - off-line programming
	Computer control and monitoring for multiple-robot/machine systems
	Standardized mechanical, electrical and computer interfaces
Assembly	Special purpose robots - small high speed Type A, medium sized Type A
	Sensory feedback systems - vision, force feedback, tactile
	Computer aided design - off-line programming
	Computer control and monitoring for multiple-robot systems
	Standardized mechanical, electrical and computer interfaces
Painting	Computer aided design - off-line programming
	Computer control and monitoring of multiple-robot systems
Inspection	Sensory feedback systems - vision, force feedback, tactile
	Computer aided design - off-line programming
	Computer control and monitoring
Material Handling	Sensory feedback systems - vision, tactile
	Standardized mechanical and electrical interfaces

Although a number of basic requirements have been identified, robotics development and limitations on it must be considered from a number of aspects. In the development area, both basic research and applications development are important. Equally important are those factors which either encourage or inhibit implementation, such as research and development, engineering support, experience and education. In all of these aspects, the United States, once a leader, appears to be losing ground to the Europeans and the Japanese, based upon the respective levels of effort in robotics development.

Although there is basic robotics research going on at more than two dozen colleges, universities and research organizations in the U.S., the programs are small, poorly funded and understaffed and largely uncoordinated. Some applications development and associated research is also carried out at research organizations and at laboratories in a few corporations such as General Motors, General Electric, Caterpillar and Westinghouse. Basic applications development is also underway in the aircraft/aerospace industry, largely as a part of the Air Force's Integrated Computer Aided Manufacturing (ICAM) Program. Although the technology developed in the ICAM Program is supposed to be generic and transferable, little of the development to date will have direct application to the automobile industry.

The National Science Foundation (NSF) funds several on-going programs, including a robot vision for randomly oriented parts ("Bin Picking") program at the University of Rhode Island (URI). Funding for this research has totalled $570 thousand over the last six years. URI is now seeking an additional $1.5 million from industry for a three year extension of this program. The Stanford Research Institute (SRI) has also been funded by NSF at a similar level for a like period for basic research in the area of Machine Intelligence. Like URI, SRI has sought industry funds for continued research in this area and for other special development programs involving sensory feedback systems for robots.

A joint development program involving Unimation, Inc., Hamilton Standard of United Technologies Corporation and Jet Propulsion Laboratory has recently been announced. The program involves the adaptation of infrared (laser) sensors for range sensing for robots and may be applicable for spot welding of car bodies on moving conveyor lines without precise orientation of the bodies. Funding amounts to $500 thousand and was provided by NASA and industry. The program is being coordinated by a joint NASA and Illinois Institute of Technology Research Institute (IITRI) organization, the NASA/IITRI Manufacturing Applications Team (MATeam).

The National Bureau of Standards (NBS) has had a modest robotics research effort underway for a number of years. The program is concentrating on sensory feedback systems (primarily vision) and control computer architecture for real-time robot servo-control from external sensors. NBS also has a machining cell with a robot being set up as part of a computer aided manufacturing development program.

In contrast to these limited (although significant) public and private development programs in the U.S., are manufacturing technology and robotics programs in Europe and Japan. Basic robotics research in Japan is being carried out in 85 laboratories in universities and public research institutions which are now working on 64 research projects. Funding for these programs is in excess of $1.5 million annually, not including salaries, wages and indirect expenses for the 350 researchers involved. The Ministry of International Trade and Industry (MITI) has recently announced a seven year, $150 million national robot research program which will begin April 1, 1982. MITI will create a new R&D group to carryout the program, which is called a "nationally important, major technology development scheme". Emphasis will be placed on intelligent robots, especially for assembly work and on robots for nuclear, space and underseas applications. The development of sensory feedback systems, programming languages and mobility will receive high priority.

The Japan Industrial Robot Association (JIRA) is also active in the administration of research programs, primarily in applications areas, such as automated systems for disposing of chips in machining processes, basic specifications of modular robots, automated system for deburring of cast iron, system design for computer aided robot system engineering and cooperation in the national project "Flexible Manufacturing System Complex with Laser". JIRA is also involved in activity to promote the development of application technology of industrial robots, administering a fund of about $3 million for this purpose.

There are also a number of long-term cooperative research programs done under government sponsorship with cooperation between specialists in the academic and industrial fields. Among these are: Industrial Robot Standardization Research Project (1974-1978); Computer Aided Robot Engineering Research Project (1976-); Pattern Information Processing Research Project (1971-1980); Casting Deburring Robotization Research Project (1975-1976, 1978-1983); Laser Applied Complex Production System Research Project (1977-1983); and Assembly Robotization Research Project (1976-1978).

There are about 140 companies manufacturing robots in Japan, including many of the largest industrial firms in the country. A number of these companies produce robots primarily for use in their own operations. The expenditures of these private enterprises on robotics development have never been announced but are estimated to far exceed the public funding. Of 107 robot manufacturers surveyed by JIRA in 1979, twenty had a specialized robot research division in their in-house research laboratories and another 52 without a special robot research division had one or more researchers specializing in basic robotic research.

In West Germany, government and industry support manufacturing technology development (which includes robotics as a significant part) at an estimated $100 million annually. The Federal Ministry for Research and Technology (BMFT) sponsors a number of programs, including "Humanization of Working Conditions", with involvement in production and manufacturing areas and robotics; development and application of manipulators, automatic handling machines and handling systems; and a new "Advanced Manufacturing Technology" program which will encompass the maintenance and increase of the competitiveness of both equipment and user companies in manufacturing technology, safety of work sites, savings of raw material and energy, reduction of environmental stresses, improvement of working conditions and the development of new handling systems. The BMFT also funds the Association for Fundamental Technology (GFK) which is involved in development programs in computer aided design and process control by computer, and the German Institute for Aerospace Research and Experimentation (DFVLR) which coordinates and monitors "Humanization of Working Conditions" and "Advanced Manufacturing Technology" programs as well as developing sensors and feedback systems for robots.

German Research Society (DFG) supports research on machine tools and controls, robotics, production systems, production engineering and manufacturing technology at the Technical Universities of Aachen, Berlin and Stuttgart. About $30 million a year is provided for these programs, half from the federal government and half from the States. Other government agencies, as well as industry, also contribute to these programs. The Association of Industrial Research Groups (AIF) is an organization that supervises the members' interests in the field of industrial research. The AIF has a budget of $100 million, which is closely matched by members' funding for individual research projects. The Fraunhofer Foundation (FhG) is an organization of institutes for applied research which performs contract research work. Two of the organization's members, the Institute for Production and Automation (IPA), Stuttgart, and the Institute for Data Processing in Technology and Biology (IITB) Karlsruhe, are current involved in industrial robot developments.

In France, the Association Francaise de Robotique Industrielle (AFRI) was founded in 1977. Within two years, it was comprised of 15 industrial groups and 8 research organizations, including both national and private research laboratories. The principle areas of research are visual and tactile pattern recognition, integrated control of robots and development of end-effectors.

The French automobile manufacturer, Renault, is deeply involved in basic robotics and in application development. Renault invested more than $15 million in a six year robot development program and continues to spend more than $2 million a year on robot research and development. Renault robot technology is considered some of the most advanced available and its ongoing R&D effort is one of the largest underway in the world.

In the United Kingdom, aside from development activities at various robot manufacturers, the major research activities are undertaken either by the government's National Engineering Laboratory (NEL) and Harwell Research Station or by academic institutions. Twelve institutions are presently involved with projects related to industrial robots. Nottingham University and Hull University are engaged in projects ranging from vision systems to computer graphics simulation of robot systems. Surrey University is working on a pneumatic powered robot and micro-processor controls. Visual feedback and computer control of robots is being investigated at Aberystwyth University. Warwick University is working on mobile robots, tactile sensors, micro-processor controls and software.

The development of ultrasonic sensors for industrial robots is being investigated at Keele University. University College, London has developed a robust, low-cost, 3-axis manipulator and research is continuing into digital control techniques needed for interaction of the manipulator with moving objects. Manipulator dynamics are also being studied at Newcastle University and Wolverhampton Polytechnic. Edinburgh University is involved in major computer software development, directed towards languages for instructing assembly robots. Queen Mary College, London, is working on computer software for control of mobile robots and for vision systems using digital array processors.

In the area of robot applications, the development of computer control for arc welding with robots is being actively pursued at NEL. The application to small batch welding is currently under investigation. At Birmingham University, a feasibility study into the robot fettling (grinding) of castings has been completed and research on the automatic detection and handling of faulty parts is being performed.

In Sweden, institutional research on industrial robots is less concerned with the development of the robot itself than with its application and integration into different production situations and systems. Research is carried out by the Swedish Institute of Production Engineering Research (IVF), Stockholm and by the universities of Lulea and Linkoping.

IVF is involved in a general study of partly manned production, where robots are essential for continuous operation. The University of Lulea is investigating the physical, mental and social effects of robots on human labor. The University of Linkoping is working on three main problem areas of automated production systems: computer-aided process planning, process supervision and machine condition monitoring and transport and handling of parts. Industrial robots are an alternative being investigated in the latter problem area, with development underway in end-effectors, adaptive control (vision and tactile sensing) and interaction between robots and other machine units.

A commission appointed by Sweden's Ministry of Industry is investigating the effect of the use of advanced electronics, including computers, in production equipment. Part of this investigation concerns the future impact of robot technology. The commission is considering whether and in what form measures shall be taken by the government either to stimulate a greater and more effective use of this technology or to improve the means for predicting and preparing the changes that may be necessary.

In Italy, robotics research is carried out in universities, in special research centers and in industry, with very little government aid. The Institute of Electro-Technics and Electronics, Polytechnic School, Milan, is involved in a number of studies, including stepping motor performance in a robot, multi-microcomputer structure, programming language, object recognition and programming of assembly problems. The Institute of Mechanics and Machine Constructing, Polytechnic School, Milan, is studying a flexible gripping device, microcomputer control, force sensors and stepping motor actuators.

The Laboratory of Mechanical Technologies, CNR-CEMU, Cinisello Balsamo, is studying rotary joint actuation with stepping motors and a programming system for a special robot. The Laboratory of System Dynamics and Biological Electronics, CNR-LADSEB, Padua, is studying algorithms for microcomputer control of mechanical members. The Laboratory of Numerical Analysis of Signals, CNR-LAMS, Turin, and the Institute of Science of Information, University of Turin, are working on voice control/programming of robots.

The Institute of Electro-Technics, University of Genoa, is studying anthropomorphic models and object recognition. The institute of Machine-Applied Mechanics, University of Genoa, is studying the theoretical aspects of identification and optimization of kinematic models of robot structures. The Institute of Machine Technology and Design, Bologna, is studying a mechanical hand provided with a movable and adjustable grip. The Institute of Automatics, University of Rome, is studying mathematical models of an articulated arm to define control parameters and the control structure of a six-legged robot for exploration of rough land.

In industry, little of the research and development work has been publicized. An exception to this is Olivetti, which manufactures relatively small, high capability robots for assembly. Olivetti's SIGMA robot has vision system capability and adaptive force feedback for close tolerance component insertions. A computer language has also been developed for off-line programming.

The foregoing descriptions of worldwide robotics research and development efforts and accomplishments would indicate that basic knowledge is not a limitation on robotic applications. As described previously, however, practical, affordable, reliable software and hardware still needs to be developed from this basic knowledge. In addition, incentives may be needed to encourage the implementation of advanced technology robotics or, indeed, to encourage the implementation of robotics in any form.

Such incentives are already in evidence in France, the United Kingdom and in Japan. In March, 1981, the French government announced the award of $50 million in aid for investment in robotics by French industry. This program, involving loans under the government's plan for the development of strategic industries, will be distributed over the next 15 months for 10 different projects.

In the United Kingdom, the Production Engineering Research Association (PERA) has, since January, 1980, been offering a low-cost advisory service to industry in the area of robot application. PERA's Robot Advisory Service (RAS) conducts briefings for top management, operates a robot demonstration center and makes free visits to firms to help management decide if there is justification for installing robots. Following this initial assessment, PERA will provide up to 15 man-days of advisory project work on a 50% cost basis. The advisory project work includes a study of the company's work handling and materials flow problems as well as the robot's tasks. Recommendations are also made regarding the robots most suitable for the tasks and a preliminary design for an end-effector may be provided. PERA is also installing, at its facility, an interactive graphics system (developed at Nottingham University) for robot simulation, to further facilitate its advisory service.

In Japan, a robot leasing company, Japan Robot Lease (JAROL) was founded in April, 1980. The aim of JAROL is to support robot installation by small and medium-sized manufacturers and increase their productivity. Because 60% of its operating funds are financed by low cost loans from the government's Japan Development Bank, JAROL's terms are more advantageous than the ordinary leasing companies. In its first year of operation (fiscal year 1980) JAROL completed 52 leasing contracts amounting to about $57.5 million. JAROL also now offers a more flexible 2-3 year rental agreement.

Other incentives arranged by MITI include direct government low-interest loans to small and medium-sized manufacturers to encourage robot installation for automating processes dangerous to human labor and for increasing productivity. For fiscal year 1980, the government budgeted more than $25 million. MITI also permits the manufacturer who installs a robot to depreciate 12.5% of its initial purchase price in the first year in addition to taking ordinary depreciation. By installing an industrial robot, a firm can depreciate 52.5% in the first year, 12.5% plus 40% (5-year depreciation double declining).

Except for components described previously (that is, sensors, interfaces, etc., there are no foreseeable problems with component availability. Accelerated usage of robots may cause delays in delivery due to lagging growth of production capacity, but this should be only a temporary situtation. Availability of knowledgeable, experienced personnel to identify, engineer and implement robot applications is more critical. It is already a problem, one which will become more severe during the next few years.

The problem is particularly acute in the North American automobile industry. In the face of rising costs and declining sales, the industry has reduced its professional staffs, particularly in the less critical areas such as manufacturing engineering and research and development. Typically, advanced manufacturing technology programs are curtailed or cancelled during such periods of economic stress. Personnel who had been involved in industrial robot programs often found that their expertise was no longer required; the robot implementation programs were cut back, delayed or cancelled and they were either reassigned, retired or laid off or they quit. Concurrently, the use of robots in other industries continued to grow and these personnel found their qualifications in demand outside of the automobile industry.

As a result, the automobile industry now faces a shortage of experienced personnel to implement new robotics programs. Formal education programs to train robotics application engineers are virtually non-existent and the products of such programs, if they were available, would still need several years of practical experience before they became effective. In-house programs to develop the necessary expertise are also virtually non-existent and, again, such "trainees" will need experience in addition to the training. Some limited assistance is available from the manufacturers and distributors of robots and from independent robotic systems engineering firms. However, these resources will be totally inadequate to support large scale robot implementation programs in the North American automobile industry.

-3-

Robot Implementation

There is a wide variety of robots available to the U.S. automobile industry. Prices, capabilities and applications also vary widely, as shown:

TABLE 4. - ROBOTS CURRENTLY AVAILABLE TO THE U.S. AUTOMOBILE INDUSTRY

TYPE	SUPPLIER	PRODUCT	PRICE RANGE ($000's)	AUTOMOTIVE APPLICATIONS
A	Armax	Armax	85-100	Spot welding, grinding
A	ASEA	IRb6	60-62	Arc welding, polishing, deburring, assembly, inspection
		IRb60	88-100	Spot welding, grinding, machine loading
A	Bra-Con Industries	Rob-Con Pacer	45-95	Press loading, plastic molding, spot welding
A	Cincinnati-Milacron	T3	75	Spot welding, arc welding, machine loading, inspection
A	Cybotech	Type 80	150	Spot welding, arc welding, machine loading
		Type 8	50-60	Light assembly
A	General Elec.	Allegro	NA	Light assembly
A	General Numeric	GNO,GN1, GN3	40-70	Machine loading
A	KUKA	IR 601/60	120-130	Spot welding
A	Prab Robots	Versatran	45-100	Spot welding, machine loading, Press loading
A	Reis	Reis Robot	60-90	Machine loading, material handling
A	Thermwood	Series Three	30-60	Machine loading, material handling, arc welding
A	Unimation	1000	25-35	Die casting, machine loading
		2000	45-65	Machine loading, spot welding, material handling
		4000	65-85	Spot welding, machine loading, material handling
		PUMA	45-50	Light assembly
A	U.S. Robots	Maker	40-50	Light assembly

TYPE	SUPPLIER	PRODUCT	PRICE RANGE ($000's)	AUTOMOTIVE APPLICATIONS
B	Automatix	Hitachi	85	Arc welding, inspection
B	Binks	88-800	50-55	Spray painting
B	Cybotech	Painter	130	Spray painting
B	DeVilbiss	Trallfa	90-100	Spray painting
B	Graco Robotic	OM 5000	120	Spray painting
B	Hobart Brothers	Yaskawa Motoman	90	Arc welding
B	Nordson	Nordson	85-95	Spray painting
B	Shin Meiwa	PW150,PW751	80-130	Arc welding
C	Armax	Armax	40-50	Die casting, material handling, machine loading
C	ASEA	Electrolux MHU	25-40	Material handling, machine loading
C	Copperweld Robotics	AP10,AP50	15-25	Material handling, light assembly, machine loading
C	Industrial Automates	Automate	15	Material handling
C	Manca/Leitz	Manta	15-45	Material handling, machine loading
C	Mobot	Mobot	15-45	Material handling, machine loading
C	Prab Robots	Prab	25-35	Die casting, plastic molding, machine loading, material handling
C	Seiko	Model 100, 200,400,700	7-15	Material handling, light assembly
D	ISI		2-10	Press loading, machine loading
D	Livernois Automation		5-20	Press loading, machine loading
D	Sterling Detroit		20-40	Die casting, plastic molding
D	Rimrock		15-25	Die casting, plastic molding

For certain applications, the type of robot which is suitable limits the users' choices and the price is fairly constant, regardless of the robot supplier. For example, spot welding (the largest single automotive application) requires a medium to large sized Type A robot and the price range for such units is about $75 thousand to $150 thousand. Spray painting requires a Type B robot with a price of $55 thousand to $130 thousand.

Within and between these classes of application and robot, the higher price reflects higher capability. Types C and D robots, which cost significantly less than Type A robots, for instance, do not have the capabilities to perform spot welding operations or to load and unload machining lines where multiple machines and buffer storage systems may be involved. On the other hand, although a Type A robot could unload a die casting machine, a Type C or D robot could also perform such a task, at a much smaller initial investment.

The total investment in a typical robot installation in the automobile industry will be from two to four times the actual robot cost, depending upon the application. In major body assembly (spot welding) systems, the cost of fixturing, welding equipment and transfer or material handling equipment will often be at least three to four times the cost of the robots. That is, a system incorporating $500,000 worth of robots may cost, in total $1,500,000 to $2,000,000. In other robot applications, which tend toward one or two robots per installation, as typified by sub-assembly welding, machine tool loading/unloading and material handling, the total investment is generally in the range of twice the robot cost.

Table 5 shows, for a typical major body assembly system, the sources and relative magnitude of the cost elements of the system.

TABLE 5. - COST ELEMENTS OF TYPICAL MAJOR ROBOT SYSTEM FOR BODY ASSEMBLY

Element	Representative Cost ($000's)	Percent of Total System Cost
Twelve Type A Robots	850	34
System design	250	10
Welding guns, transformers and controls	150	6
Conveyors and/or transfer mechanisms	150	6
Locating and positioning fixturing	250	10
Controls and interfacing	200	8
Safety devices, guard rails, etc.	50	2
Site preparation	150	6
System assembly, tryout and shipping	250	10
System installation, robot programming and debugging	100	4
Personnel training	25	1
Efficiency and production losses during start-up	75	3
TOTAL	$2,500	100%

The sources and relative magnitude of the cost elements for a simple, one-robot installation applied to load and unloading two machine tools are shown in Table 6:

TABLE 6. - COST ELEMENTS OF TYPICAL SINGLE ROBOT INSTALLATION FOR MACHINE TENDING

Element	Representative Cost ($000's)	Percent of Total System Cost
One Type A Robot	55	44
System design	10	8
End-of-arm tooling	5	4
Conveyors and part orienters	15	12
Controls and interfacing	7	6
Safety devices, guard rails, etc.	5	4
Rearrangements and site preparation	5	4
Equipment relocation and revision	5	4
System installation, robot programming and debugging	5	4
Personnel training	3	2
Efficiency and production losses during start-up	10	8
TOTAL	$125	100%

The use of robots in automobile manufacturing operations incurs costs beyond the initial investment in the robot system. Like any machine, the robot requires periodic maintenance and is subject to occasional breakdowns, necessitating repair and/or replacement of components. The cost of robot maintenance and repair will vary widely, depending upon the type or robot, severity of the task and level of maintenance provided. For cost analysis and budget purposes, one major automobile manufacturer has developed some standards, based upon actual experience.

Using the recent maintenance records from one of its assembly plants which is operating 27 Type A robots of several brands, the cost of replacement parts and consumable supplies (hydraulic fluid, lubricants, filters, etc.,) was determined. Averaged over the number of robots and production shifts, a standard of $4.41 per 8-hour shift was established. This is about $0.55 per operating hour for maintenance and repair materials only. Incidently, this new (June, 1981) standard was a significant reduction from the $5.69 per 8-hour shift or $0.71 per operating hour standard which had been previously established in 1973.

In addition to the maintenance and repair material standard, there is a labor allowance for maintenance. This same major automobile manufacturer has established a standard of 1.698 hours per day per robot, operating on two production shifts and 1.029 hours per day per robot, operating on one production shift. Using a current indirect or maintenance labor rate of about $19.05 per hour (including fringe and related costs), this standard amounts to about $2.16 per operating hour. When the labor cost is added to the maintenance and repair material allowance, the total maintenance cost is about $2.71 per operating hour for Type A robots.

The costs for Type B robots would be about the same; this automobile manufacturer uses the same allowances for all servo-controlled robots. Maintenance and repair costs for Type C and Type D robots would be significantly less--on the order of $0.50 per hour--due to their simplicity and the much lower cost of control and feedback components. A representative hourly maintenance and repair cost for Types A and B robots would be about 0.004 percent of the initial cost of the robot. In Types C and D robots, this representative cost would be about 0.002 percent of the initial cost of the robot.

Operating costs for a robot, in addition to maintenance and repair are primarily those related to "consumables" required by the robot to perform its task. These would include electrical energy or compressed air to power the robot and its control and, perhaps, compressed air to actuate a gripper or tool and, for some installations, water for cooling of hydraulic fluid and/or the control console.

In many applications, these "consumable" requirements are partially or completely offset by reduced requirements for lighting, heating, cooling or ventilation formerly needed by the human worker and, perhaps, reduced energy consumption by the process or machine serviced by the robot as a result of the robot's higher efficiency and productivity.

Other operation costs include amortization of investment, taxes, insurance, etc.

Unimation Inc. uses an hourly operation cost for their robots (which are servo-controlled, Type A) of about $5.00 per operating hour. This cost includes maintenance, as well as operation, and is based upon an average robot purchase price of about $60,000. Cincinnati-Milacron uses a cost of $6.00 per operating hour, which reflects the somewhat higher average cost of their robot (compared to a Unimate). A representative total hourly operation cost, including all of the elements described above, would be about 0.008 percent of the initial cost of the robot for Types A and B. For Types C and D robots, the total hourly operation cost would be about 0.005 percent of the initial robot cost.

-4-

Present Status of Robots in Worldwide Automobile Manufacturing

At present, North America, Asia and Western Europe are the three major automobile manufacturing regions. Each of these geographical areas produces roughly the same quantity of automobiles, as indicated in Table 7, which shows world passenger car production for 1980. During the first three months of 1981, U.S. and Western Europe automobile production declined somewhat, compared to the first quarter of 1980, while Japanese production increased slightly, bringing regional production even more into balance.

TABLE 7. - 1980 WORLD PASSENGER CAR PRODUCTION AND ASSEMBLY-SELECTED COUNTRIES

Region	Country		1980 Production
North America	United States		6,373,071
	Canada		844,973
		Sub Total	7,218,044
Western Europe	West Germany		3,306,031
	France		2,714,466
	Italy		1,329,079
	United Kingdom		924,100
	Belgium		676,920
	Spain		953,988
	Sweden		197,248
		Sub Total	10,101,832
Asia	Japan		7,038,108
South America	Argentina		172,867
		GRAND TOTAL	24,530,851

TABLE 7. (continued)

Sources: Ward's Statistical Department from various world sources. In some of the lower volume countries for 1980 the 12-month totals are estimates based on 10- and 11-month actual counts. (Ward's Auto World, March, 1981)

The availability of technology and extent of utilization is also distributed fairly uniformly within each of these geographical areas. That is, technology in West Germany, for example, is quite similar to that in France and Italy, but not necessarily similar to that in Japan or the U.S. It is convenient, therefore, to compare geographical areas when establishing some overall positions relative to utilization of automation and industrial robots.

Table 8 shows the estimated population of industrial robots at year-end 1980 in fifteen countries in Asia, North America, Western Europe and Eastern Europe.

TABLE 8 - ESTIMATED INDUSTRIAL ROBOT POPULATION AT YEAR-END 1980

Region	Country	Robot Population by Type a/					Total
		A	B	C	D	E	
North America	United States	2485	439	1377	*	*	4,301
	Canada	54	5	10	*	*	69
						S/T	4,370
West Europe	West German	440	293	200	200	5000	6,133
	Sweden	720	180	300	*	*	1,200
	Italy	300	50	100	50	*	500
	France	320	60	150	*	*	530
	Norway	20	20	100	30	30	200
	United Kingdom	146	117	54	54	*	371
	Finland	20	10	70	10	20	130
	Belgium	42		82	3	7	134
	Spain	29	*	*	*	*	29
						S/T	9,227
East Europe	Poland	70	40	200	50	360	720
	USSR	100	50	200	150	2000	2,500
						S/T	3,220
Asia	Japan	5150		12050		50510	67,710
	Australia	14	*	*	*	*	14
						S/T	67,724
				GRAND TOTAL			84,541

*Data Unavailable

a/ Type A: Programmable, Servo-Controlled, Point-to-Point
Type B: Programmable, Servo-Controlled, Continuous-Path
Type C: Programmable, Non-Servo, General-Purpose
Type D: Programmable, Non-Servo, Special-Purpose for Die Casting and Plastic Molding Machines
Type E: Non-Programmable, Mechanical Transfer Devices

Sources: Robot Institute of America, March, 1980, Daiwa Securities America Inc., July, 1981 and others

Although all of the robot types shown in Table 8 are used by automobile manufacturers or their suppliers, only the programmable devices (Types A, B, C and D) are of significance as indicators of the extent and level of automation technology utilization. A more useful comparison of technology levels is shown in Table 9, which summarizes the Types A through D robot populations by region.

It should be noted that these estimates are based upon reported or published data and do not include figures from several Western European countries or from most Eastern Europe countries, although it is known that robots are being used in those countries.

TABLE 9. - ESTIMATED PROGRAMMABLE ROBOT POPULATION AT YEAR-END 1980, BY REGION

Region	Total Robot Population (Types A, B, C and D)
North America	4,370
Western Europe	4,170
Eastern Europe	860
Asia	17,214

The automobile industry, that is, the automobile manufacturers and their suppliers, is the largest single industry using robots today. Table 10 shows the number of robots in use in 1980 in the automobile industry, by region. Also shown is the percentage of the total robot population so used.

TABLE 10 - ESTIMATED PROGRAMMABLE ROBOT USAGE IN THE AUTOMOBILE INDUSTRY AT YEAR-END 1980

Region	Automotive Robots (Types A,B,C and D)	Percent of Total Robots
North America	1,065	24%
Western Europe	1,688	40%
Asia	5,217	30%

Sources: Robot Institute of America, Japan Industrial Robot Association, National Robot Coordinators Reports, March, 1979 and March, 1980, Daiwa Securities America Inc., July, 1981, and others

A detailed breakdown, by automobile companies and suppliers, of the year-end 1980 automotive industry robot population is shown in Table 11. Data for this Table was derived from a variety of sources; magazine articles, internal reports, robot manufacturers, etc., and represents the best estimates of the current situation.

TABLE 11 - ESTIMATED PROGRAMMABLE ROBOT USAGE BY AUTOMOBILE MANUFACTURERS AND SUPPLIERS AT YEAR-END 1980

Region	Manufacturer/Supplier (Country)	Number of Robots A	B	C	D	TOTAL
North America	American Motors (U.S.)	6	-	4	-	10
	Chrysler (U.S.)	210	-	-	-	210
	Ford (U.S./Canada)	258	12	5	-	275
	General Motors (U.S./Canada)	389	36	25	-	450
	Suppliers (U.S./Canada)	60	10	50	-	120
	Sub Total	923	58	84	-	1,065
Western Europe	Alfa Romeo (Italy)	28	2	-	-	30
	British Leyland (U.K.)	34	6	-	-	40
	BMW (West Germany)	150	11	-	-	161
	Daimler Benz (West Germany)	40	51	-	-	91
	Fiat (Italy)	240	10	-	-	250
	Ford-Europe (U.K., W.Germany Belgium, Spain)	105	8	2	-	115
	Opel/Vauxhall-GM (W.Ger., U.K)	100	16	-	-	116
	Peugeot/Citroen/Chrysler (France, U.K.)	30	12	-	-	42
	Renault (France)	178	36	-	-	214
	Rolls Royce (U.K.)	2	1	-	-	3
	Saab (Sweden)	25	8	19	-	52
	Volkswagen (West Germany)	192	8	20	-	220
	Volvo (Sweden)	150	13	45	-	208
	Suppliers - Italy	8	12	5	-	25
	- France	10	2	6	-	18
	- Sweden	5	15	10	-	30
	- U.K.	7	3	-	13	23
	- West Germany	10	5	35	-	50
	Sub Total	1,314	219	142	13	1,688

TABLE 11 (continued)

Region	Manufacturer/Supplier (Country)	Number of Robots A	B	C	D	TOTAL
Asia	Ford (Australia)	14	-	-	-	14
	Fuji/Subaru (Japan)	50	15	5	-	70
	Honda (Japan)	100	16	10	-	126
	Isuzu (Japan)	2	-	-	-	2
	Mitsubishi (Japan)	250	20	750	-	1,020
	Nissan (Japan)	320	40	930	30	1,320
	Suzuki (Japan)	90	10	130	-	230
	Toyo Kogyo (Japan)	275	10	150	-	435
	Toyota (Japan)	420	30	850	-	1,300
	Suppliers (Japan)	110	60	470	60	700
	Sub Total	1,631	201	3,295	90	5,217
	Grand Total	3,868	478	3,521	103	7,970

Automobile manufacturing operations can be separated into several uniquely different groups, to which varied levels and types of automation are applied. These groups include:

- Component Manufacturing and Assembly
- Body Assembly
- Body Finishing
- Trim and Final Assembly

Each of these groups of manufacturing operations and the levels and types of automation utilized are described on the following pages.

-5-

Current Applications in Component Manufacturing and Assembly

Automobile component manufacturing and assembly covers a wide variety of processes, operations and materials. Typical automobile components where automation is employed to a significant degree include engines, transmissions, axles, sheet metal and plastic body panels, frame and suspension members, steering assemblies and electrical and electronics systems.

The processes involved include casting and forging, component (small parts) machining, pallet-transfer machining, stamping and molding, and assembly and testing. No particular automobile manufacturer or region appears to have any advantage in the automation of these processes; the technology is widespread and available to all.

Types A, B and C robots are used for a variety of operations associated with component manufacturing and assembly, handling either parts or tools.

In the foundries and forging shops, Type A, B, C and D robots are used in a variety of operations. New casting techniques and forming processes are also being employed. Examples of these technologies are shown in Table 12.

TABLE 12 - FOUNDRY AND FORGING TECHNOLOGY UTILIZATION

Region	Company (Country)	Technology
North America	Doeler-Jarvis (U.S.)	Unload die casting machines with robots, handling parts up to 75 pounds, about 50 robots in use.
		3000 Ton capacity die casting machine with robot unloader to produce parts for G.M. Hydra-matic
	Federal-Mogul (U.S.)	Hot forming of powder metal preforms for transmission components and bearing races.
	Ford (U.S.)	Unload die casting machines with robots, 8 robots in use.
	General Motors (U.S.)	Automatic pouring of V-6 cast iron engine blocks.
		Hot forming of powder metal preforms for transmission components.
		Unload die casting machines with robots, 7 robots in use.
		Flaskless (continuous, automatic) moulding lines for cast iron parts, robots transfer cast parts into die triming presses, 5 robots in use.
		New automatic rotary casting equipment installed in Central Foundry.
		Seven fully automatic forging presses being installed at Chevrolet Detroit Forge plant.
Western Europe	Alfa Romeo (Italy)	Robots load hot blanks into forging press and forged parts into triming press, 2 robots in use.
		Unload die casting machines with robots, 2 robots in use.
		Remove gates and risers and grind flash from castings with a robot.

TABLE 12 (continued)

Region	Company (Country)	Technology
W.Europe (Cont'd)	British Leyland (U.K.)	Grind flash from cylinder heads and engine blocks with a robot.
	Castings (U.K.)	Flaskless (continuous, automatic) moulding line for small malleable iron parts such as door hinge and spring brackets.
	Citroen (France)	Robot positions cores in molds.
	Daimler Benz (West Germany)	High-pressure water jet system to decore (remove sand from internal passages) aluminum cylinder blocks.
		Robot handling forgings at Unterturkheim Plant.
	Fiat (Italy)	Unload die casting machines with robots, 12 robots in use.
	Ford (West Germany)	Robot handling forgings at Koln plant.
	Metal Castings (U.K.)	Unload die casting machines with robots, 21 robots in use.
		Robot installed on 600 ton die caster to load inserts and unload aluminum automobile parts.
	Renault (France)	Automated cylinder liner machines (complete from pouring of metal to removing castings and placing them in baskets on a conveyor).
	Tube Investments (U.K.)	Robot handles axle components through upsetting (forging) operations.
	Volkswagen (West Germany)	Grind flash from engine blocks with a robot.
		Robot transfers cast crankshafts from conveyor to containers.

TABLE 12 (continued)

Region	Company (Country)	Technology
W.Europe (Continued)	Volvo (Sweden)	Grind flash from gear box castings with a robot.
		Clean molds with robot handling air jet.
		Load billets into forging press with a robot.
		Continuous path (Type B) robot spraying cores and molds in foundry.
Asia	Nippondenso (Japan)	Sixty robots (Types A and B and C) unloading die casting machines.
	Nissan (Japan)	Unload die casting machines with robots, 20 robots in use.
		High-speed automatic forging of gear blanks at 35 to 70 per minute.

In basic parts manufacturing, the new technologies in use include finish rolling of gears, laser and electron beam welding and hardening, Types A and C robot parts handling and a variety of inspection techniques. Examples are shown in Table 13.

TABLE 13 - BASIC PARTS MANUFACTURING AND INSPECTION TECHNOLOGY UTILIZATION

Region	Company (Country)	Technology
North America	Chrysler (U.S.)	Electron beam welding of two-piece die cast aluminum intake manifold, with computer-controlled part positioning.
	Ford (U.S.)	Automatic camshaft gaging system with automatic parts handling and computer interface.
		Transmission gear finish rolling machines, 3 machines.
	General Motors (U.S.)	Automatic handling and inspection system for pistons (2,000 per hour).
		Automatic handling and inspection system for valve rocker arms (9,000 per hour).
		Laser welder for Packard Electric Division to produce wire-to-terminal assemblies for on-board computer-control systems, to join wire made of unconventional materials to terminals at high production rates.
		Five laser welders for AC Spark Plug Division to assemble exhaust oxygen sensors for computer-controlled catalytic converter system.

TABLE 13 (continued)

Region	Company (Country)	Technology
Western Europe	British Leyland (U.K.)	Transmission gear finish rolling machines, 40 machines in place or on order.
	Borg-Warner (U.K.)	Transmission gear finish rolling machine.
	Fiat (Italy)	Laser welding of synchronizer rings to gears.
		Laser hardening of crankshafts.
		Laser surface hardening of gear selectors.
	Ford (U.K. & W.Germany)	Transmission gear finish rolling machines, 6 lines.
	Ford (U.K)	Two robots deburring machined suspension components.
	Ford (W.Germany)	Two robots deburring machined components.
	Kongsberg (Norway)	Highly-automated production line for axle half-shafts for volvo.
	Lancia (Italy)	Automatic gear cutting machines with automatic parts feeding.
		Electron beam welding of synchronizer rings to gears.
	Saab (Sweden)	Robots loading and unloading metal-cutting machines, 15 robots in use with lathes, broaches, drilling machines, gear shapers and milling machines.
	Volvo (Sweden)	Three robots handling and grinding components.
		In crankshaft department at Skovde Engine Plant, 40 of the 60 machine tools are equipped with dedicated robots to load and unload parts.

TABLE 13 (continued)

Region	Company (Country)	Technology
Asia	Nissan (Japan)	Fixed-sequence manipulators load and unload gear turning and cutting machines, 1,500 manipulators in use in the Yoshiwara Transmission Plant.

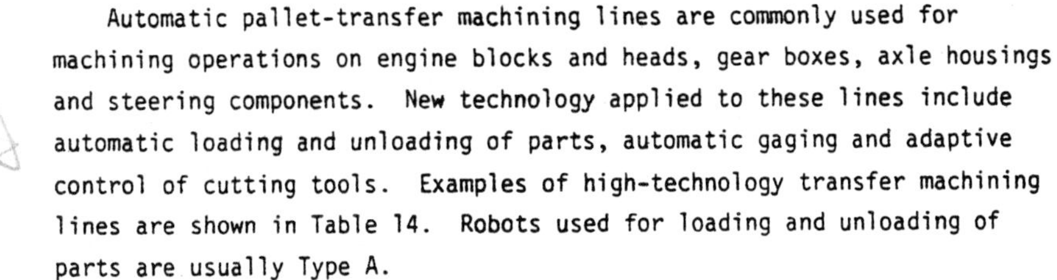

Automatic pallet-transfer machining lines are commonly used for machining operations on engine blocks and heads, gear boxes, axle housings and steering components. New technology applied to these lines include automatic loading and unloading of parts, automatic gaging and adaptive control of cutting tools. Examples of high-technology transfer machining lines are shown in Table 14. Robots used for loading and unloading of parts are usually Type A.

TABLE 14 - HIGH-TECHNOLOGY TRANSFER MACHINING LINE UTILIZATION

Region	Company (Country)	Technology
North America	Ford (U.S.)	Multi-station transfer line for differential case machining, incorporating four CNC turning stations to perform 3-axis contouring; in-line gaging and automatic tool compensation.
		Large line to machine aluminum transmission extension housings, utilizing diamond tooling.
		Large Japanese transfer line to manufacture lock-up torque converter housings for automatic transmissions.
		46-Station transfer line to machine power steering rack pistons, with display of part clamping torque and in-line pallet inspection station.
	General Motors (U.S.)	11-Machine integrated manufacturing system to machine, assemble and test rear axles.
		37 Robots loading and unloading transmission case pallet transfer machining lines.
		Large in-line system to machine aluminum intake manifolds, using high-water-based fluid for coolant and transfer drive.

TABLE 14 (continued)

Region	Company (Country)	Technology
N.America (Cont'd)	General Motors (U.S.)	Incorporation of laser inspection systems in a valve lifter production line to inspect blanks before machining.
		40-Station line for machining clutch housings and 37-station line for transmission cases, with high-accuracy drives and bore gaging systems, which permit machining housing and case halves separately, instead of as assemblies.
		Programmable transfer lines for engine component machining at Cadillac and Oldsmobile.
		Flexible machining system -- NC machining centers and computer-controlled transporters being purchased from Comau (Fiat) for Chevrolet Gear and Axle.
		New high-speed pallet transfer line for machining steering knuckles at rate up to 200 sets (left-hand and right-hand) per hour, at Chevrolet Saginaw Manufacturing Plant.
Western Europe	BMW (West Germany)	Engine block machining line with in-line bore gaging and automatic tool adjustment.
		Crankshaft machining line with in-line heat treating and automatic 100% inspection of critical dimensions.
		Cylinder head machining line with robot deburring heads after machining.
	Lancia (Italy)	Transmission case machining line which handles five different cases.

TABLE 14 (continued)

Region	Company (Country)	Technology
W.Europe (Cont'd)	Opel (West Germany)	Three 45-station engine block machining lines, for Australia, Brazil and Germany. Australian line will handle seven different blocks.
	Volvo (Sweden)	Four robots deburr gear box parts after machining.
		High-volume, three section transfer line for machining steering knuckles at rate up to 114 pairs per hour. Totally automatic, including loading, machining, gaging and tool adjustment.

Metal stamping and forming and plastics forming operations have been automated, to some degree, for some time. Mechanical transfer systems, non-programmable manipulators and Types A, B and C robots are used for these tasks. Some examples of recent innovations in these operations are shown in Table 15.

TABLE 15 - METAL AND PLASTICS FORMING TECHNOLOGY UTILIZATION

Region	Company (Country)	Technology
North America	A. O. Smith (U.S.)	Five robots used on three press lines to transfer frame rails and cross-members between stamping presses at 340 to 390 parts per hour.
	Bendix (U.S.)	In-line stamping system from Japan with automatic coil feeding, blanking. blank handling and stamping sections and a powered movable bolster for changing dies; for front suspension members.
	Ford (U.S.)	Four robots arc welding door frames.
	General Motors (U.S.)	Automated press and welding line to form and weld fender reinforcements, pierce and install clinch nuts at a rate of 1,440 parts per hour.
		Integrated high-speed transfer line to assemble, weld and hem flanges on aluminum and steel hoods.
		Automatic high-speed inspection system for engine valve covers with automatic leak tester, solid-state camera visual inspection, programmable controller and robot handling, at a rate of 1,200 per hour.

TABLE 15 (continued)

Region	Company (Country)	Technology
N.America (Cont'd)	General Motors (U.S.)	Three robots arc welding components at Delco Plant. Four robots welding J-Car trailing axles at Pontiac. Twenty-six robots loading and unloading stamping presses at Fisher Body Plants. Five robots arc welding in Bedford Foundry.
		Gamma-ray parts-detector system checking J-Car welded sheet metal assemblies for missing components.
	Hayes-Albion (U.S.)	14-Station automatic line for grinding and polishing weld joints of roll-formed and welded door glass frames, with automatic wheel wear compensation.
Western Europe	Alfa Romeo (Italy)	Transfer devices between large presses; large, coil-fed presses with transfer dies in place of series of small presses; gravity feeders on small presses; non-programmable pick-and-place devices to load and unload presses.
	Armstrong Patents (U.K.)	Computer-controlled vector bending machines for forming exhaust pipes; automatic assembly of mufflers; automatic welding of exhaust pipes; mechanical locking of pipes to mufflers instead of welding; automatic forming and welding of mufflers.
	British Leyland (U.K.)	Three robots arc welding front sub-frames and other components at Longbridge Plant.
	BMW (West Germany)	Nine robots arc welding components at Dingolfing and Munich Plants.
	Daimler Benz (West Germany)	Seven robots arc welding components at Unterturkheim, Mannheim and Sindelfingen Plants.

TABLE 15 (Continued)

Region	Company (Country)	Technology
W.Europe (Cont'd)	Fiat (Italy)	Automatic press-to-press transfer systems, including automatic loading and unloading of dies.
		Five robots arc welding components.
	Ford (U.K.)	6-Station press line for one-piece body sides, with non-programmable manipulators for loading and unloading and automatic press-to-press transfer.
	Ford (West Germany)	One robot arc welding components at Koln. One robot loading and unloading stamping press at Koln.
	Opel (West Germany)	Sixteen robots arc welding front suspension and rear axle components.
	Peugeot (France)	Four robots arc welding components at Peugeot and Citroen Plants.
	Renault (France)	Automatic loaders and unloaders for stamping presses; 6-press line for door frames with automatic loading, transfer and unloading at 600 parts per hour; automatic bumper polishing and plating lines.
		Ten robots arc welding components.
	Saab (Sweden)	Seven robots arc welding components.
	Tube Investments (U.K.)	One robot arc welding rear axle housings.
	Volvo (Sweden)	Four robots arc welding components.

TABLE 15 (Continued)

Region	Company (Country)	Technology
Asia	Fuji (Japan)	Thirteen robots arc welding rear suspension crossmembers and rear suspension arms.
	Honda (Japan)	Seven-press bodyside stamping line at Sayama Plant with robots to load and unload presses, operates on 5-second cycle.
	Mitsubishi (Japan)	Three robots arc weld brackets to rear axle housings.
	Nissan (Japan)	Four robots arc weld engine mounting brackets and seat frames.

Component assembly and testing operations have often been highly labor intensive, with automation used primarily for transfer of parts between manual work stations. The recent trend is to incorporate automation for assembly, inspection and testing, as well as for parts handling, as shown in examples in Table 16. Types A and C robots are used in these operations.

TABLE 16 - AUTOMATED ASSEMBLY AND TESTING UTILIZATION

Region	Company (Country)	Technology
North America	Chrysler (U.S.)	Semi-automatic, non-synchronous carousel transfer line for trans-axle assembly. Twenty-four station line produces 183 assemblies per hour at 100% efficiency.
	Ford (U.S.)	Laser gage to inspect for washers on gear carriers; non-contact gage to verify rear axle gear ratios.
		Computerized engine hot test system with 38 test stands, 4 diagnostic stands and 6 repair stations, served by automatic non-synchronous engine handling system.
		Two 65-station, 3-loop systems to automatically assemble, inspect and test steering gear rack and pinion assemblies at rates of 240 per hour.
		156-Station, 3-loop system to automatically assemble, inspect and test front-wheel-drive transaxle assemblies.
	General Motors (U.S.)	Automatic assembly machine for flow control valves for power steering pumps.
		Five automatic assembly and welding systems for catalytic converters, with robots packing finished products into shipping containers.

TABLE 16 (Continued)

Region	Company (Country)	Technology
N.America (Cont'd)	General Motors (U.S.)	Eight robots assembling air conditioner blower motors; one robot assembling motor commutators to armatures; one robot installing screws in carburetors.
		Several transmission assembly systems with automatic parts rejection, functional testing, size gaging,shim selection and bolt feeding and torquing.
		Automated hot- and cold-test systems for 1.8 liter and 4.1 liter engines.
		One robot inserting wear sensor in brake shoes. One robot palletizing magnet assemblies for radio speakers.
		Computerized vision systems inspecting head assemblies for keepers in valve springs.
		Laser inspection system checks surfaces of valve lifter bodies.
Western Europe	Alfa Romeo (Italy)	Robot deburrs engine front cover, inserts bushing and drives studs.
	British Leyland (U.K.)	11-Station transfer machine to install valve guides and valve seat inserts in aluminum cylinder heads.
		14-Spindle automatic bolt tightener with torque/angle control for cylinder head-to-block assembly.
	BMW (West Germany)	Automated pallet-transfer engine assembly line, with all lifting and fastening mechanized; automated cylinder head assembly line, both with computer monitoring; about 60% of operations are automated.
	Fiat (Italy)	Robot assembly and torque testing of steering gear tie rod assemblies; semi-automatic assembly of engines, using robots.

TABLE 16 (Continued)

Region	Company (Country)	Technology
Western Europe	Ford (West Germany)	Robot loads and unloads automatic leak testing machine for transmission cases.
	Volvo (Sweden)	New engine manufacturing plant, with self-propelled, wire-guided, computer-controlled carriers to transport engines through assembly area, "group" or "team" assembly concept, computerized, automatic hot testing (including emissions tests) after assembly.
Asia	Nippondenso (Japan)	High speed automatic assembly lines for automotive instrument assembly. Fuel gage assembly line handles 60 different assemblies in small batches (as many as 200 batches per day), operates on a one-second cycle and produces 500,000 per month on a single shift.
	Nissan (Japan)	Two pallet-transfer transmission assembly lines, with 60% of operations automated, including component insertion and fastener tightening.

-6-

Current Applications in Body Assembly

Automobile body assembly operations involve resistance (spot) welding of body panels into major subassemblies, welding of these subassemblies into a body shell, metal finishing, dimensional inspection and mechanical installation of components such as doors, hoods and trunk lids. It is in the welding operations that the highest degree of automation is to be found. Two approaches to automated welding are used, special-purpose multiwelders and Type A robots and most manufacturers utilize a combination of both technologies.

Table 17 shows the extent of automation of welding operations of the major automobile manufacturers. Also shown are the numbers of industrial robots used in welding operations. Data for this Table was derived from a variety of sources; magazine articles, internal reports, robot manufacturers, etc. and represents the best estimates of the current situation.

TABLE 17 - ESTIMATED PERCENTAGE OF BODY WELDING DONE WITH AUTOMATION (MULTIWELDERS AND ROBOTS) AND NUMBER OF ROBOTS WELDING AT YEAR-END 1980

Region	Manufacturer (Country)	(Percent) Automated Welding	(Type A) Number of Robots Welding
North America	American Motors (U.S.)	25	5
	Chrysler (U.S.)	75	206
	Ford	60	240
	General Motors (U.S./Canada)	45	220
	Regional Average	52%	S/T 671
Western Europe	Alfa Romeo (Italy)	40	10
	British Leyland (U.K.)	60	30
	BMW (West Germany)	85	146
	Daimler Benz (West Germany)	80	40
	Fiat (Italy)	75	220

TABLE 17 (Continued)

Region	Manufacturer (Country)	(Percent) Automated Welding		(Type A) Number of Robots Welding
W.Europe (Cont'd)	Ford-Europe (U.K., West Germany, Belgium, Spain)	45		100
	Opel/Vauxhall-GM (West Germany, U.K.)	60		100
	Peugeot/Citroen/Chrysler (France, U.K.)	50		24
	Renault (France)	90		178
	Rolls Royce (U.K.)	0		0
	Saab (Sweden)	85		80
	Volkswagen (West Germany)	95		150
	Volvo (Sweden)	95		150
	Regional Average	72%	S/T	1228
Asia	Ford (Australia)	25		2
	Fuji/Subaru (Japan)	85		20
	Honda (Japan)	80		81
	Isuzu (Japan)	40		2
	Mitsubishi (Japan)	80		200
	Nissan (Japan)	95		320
	Suzuki (Japan)	60		80
	Toyo Kogyo (Japan)	80		250
	Toyota (Japan)	85		300
	Regional Average	88%	S/T	1255
	Overall Average	70%	Grand Total	3154

Table 18 shows some examples of body assembly technology which are noteworthy because of either the extent of automation or the uniqueness of its application.

TABLE 18 - AUTOMATION TECHNOLOGY UTILIZATION IN BODY ASSEMBLY

Region	Company (Country)	Technology
North America	Chrysler (U.S.)	Bevidere, Illinios plant uses Robogate (Fiat) system with 50 robots to weld Omni and Horizon body shells -- about 90% of welding is automated.
		Detroit (Jefferson) and Newark, Delaware plants use Robogate (Fiat) systems and total of 128 robots to weld K-Car body shells; about 98% of welding is automated.
		St. Louis Plant uses indexing body welding line and a total of 14 robots to weld body shells.
	Ford (U.S.)	Four robots with contact probes, programmable controller and computer for data analysis check 150 dimensions on car bodies at a rate of 5 to 6 per hour.
	General Motors (U.S.)	Lakewood, Georgia plant uses four robots on continuous-moving line to weld body shells.
		Lordstown, Ohio and South Gate, California plants each use 36 robots to weld body shells and two robots with laser gages for dimensional checking of bodies after welding.
		Willow Run, Michigan plant uses two robots with force feedback to wire brush weld joints, after welding, before painting.

TABLE 18 (Continued)

Region	Company (Country)	Technology
Western Europe	British Leyland (U.K.)	Longbridge plant uses two four-stage multiwelders for underbody welding, two three-stage multiwelders and 28 robots, on two lines to weld ADO 88 body shells. Body shell welding incoroporates checking fixtures after last multiwelder to check 22 critical dimensions before robot welding. About 98% of welding is automated.
	Daimler Benz (West Germany)	Sindelfingen plant has most highly mechanized body assembly operations in Europe, with multiwelders and robots performing virtually all welding automatically.
	Fiat (Italy)	Mirafiori plant has 3-stage multi-welder with automatic checking station after and 23 robots welding Model 131 bodies, multiwelder and 18 robots welding Model 132 bodies.
		Cassino plant has 2-stage multiwelder for Model 126 bodies, 45% of welding is automatic; multiwelders for Model 128 bodies, 63% of welding is automatic; multiwelders for Model 127 bodies, 83% of welding is automatic; multiwelders for Strada underbodies and Robogates for bodysides and body shells with 57 robots, 60% of welding is automatic.
	Ford (U.K./West Germany)	Halewood plant has multiwelders and 39 robots for underbodies, bodysides and body shells, about 90% of welding is automatic.
		Saarlouis plant has multiwelders and robots for underbodies, bodysides and body shells.
		Cologne plant has 25 robots spot welding body shells.

TABLE 18 (Continued)

Region	Company (Country)	Technology
W.Europe (Cont'd)	Peugeot (France)	3-Stage multiwelder and 15 robots weld body shell.
	Renault (France)	Flins plant has multiwelders for R5 subassemblies, underbody and body shells, 48% of welds are automatic; multiwelders for R12 subassemblies, underbody and body shell, 46% of welds are automatic; multiwelders and 12 robots for R18 subassemblies, underbody, bodysides and body shell, 82% of welds are automatic.
		Douai plant has multiwelders and 96 robots for R5 and R14 subassembly, underbody, bodyside and body shell welding.
	Saab (Sweden)	Trollhattan plant has multiwelders and 10 robots for Model 900 subassembly, underbody, bodyside and body shell welding, 98% of welding is automatic.
	Volvo (Sweden)	Torslanda plant has multiwelders and 46 robots for subassembly, underbody bodyside and body shell welding, 95% of welding is automatic. Two robots with contact probes and computer interface check body dimensions after robot welding.
Asia	Fuji Heavy Industries (Japan)	Gunma plant has multiwelders and seven robots for subassembly, underbody, bodyside and body shell welding for Datsun and Subaru body welding, 89% of welding is automated.
	Honda (Japan)	At Suzuka plant, 80 robots and several special-purpose multiwelders fit and weld body shells. More than 80% of welding is automatic.

TABLE 18 (Continued)

Region	Company (Country)	Technology
Asia (Cont'd)	Mitsubishi (Japan)	Okazaki plant has multiwelders and 77 robots for subassembly, underbody, bodyside and body shell welding, about 80% of welding is automated; at Mizushima plant, 70 robots are used for bodyside and underbody welding; at Oihye plant, 35 robots are used for welding.
	Nissan (Japan)	Zama plant has 13 lines, 49 multiwelders and more than 50 robots to weld subassemblies, bodysides, underbodies and body shells, about 96% of welds are made automatically; Oppama plant has multiwelders and 64 robots for subassembly, underbody, bodyside and body shell welding; Mirayama plant has multiwelders and 85 robots for subassembly, underbody, bodyside and body shell welding, 63% of welds are made automatically.
	Nissan (Japan)	At Tochigi plant, 6 Kawasaki 6060 arms in one station weld four different models of side assemblies; each robot arm makes 8 welds in 20 seconds. This station is followed by four more stations with 10 Kawasaki 2000 robots.
	Toyo Kogyo (Japan)	77 Robots have been installed in body shop for assembly of new small family car, the Familia or 323; previously, about 30 robots were being used. New plant at Kifu has 165 robots for Mazda cars.
	Toyota (Japan)	Takoaka plant has multiwelders and robots for subassembly, underbody, bodyside and body shell welding; New Tahara plant will have about 200 robots for welding installed in 1981.

-7-

Current Applications in Body Finishing

Body finishing generally combines automation and manual operations, to varying degrees. After surface preparation (washing, phosphate treatment), most automobile manufacturers apply the first prime coat by electro-deposition in a dip tank. Both of these steps are automated. Final prime and color applications are sprayed.

Spray painting generally combines automatic equipment to cover the major exterior surfaces and hand "cut-in" spraying of interiors, door edges, underside of hoods and trunk lids, etc. Protective materials are applied to the underside of wheel openings, etc., for corrosion prevention and sound deadening, after other finishing operations are complete.

The extent of automation for final prime and color application and for undersealing is higher in Western Europe and Japan than in North America, with Type B robots used in place of manual spraying as shown in Table 19.

TABLE 19 - AUTOMATION TECHNOLOGY UTILIZATION IN BODY FINISHING

Region	Company (Country)	Technology
North America	American Motors (U.S.)	Stationary spray gun, electrostatic primer application, with manual "cut-in".
	General Motors (U.S.)	Stationary spray gun, electrostatic primer, sealer and color application, with manual "cut-in"; four paint lines, with four robots in each line, to spray primer and finish color on urethane plastic bumpers.
	Volkswagen (U.S.)	Automatic electrostatic primer and color coating, with manual interior spray.

TABLE 19 (Continued)

Region	Company (Country)	Technology
Western Europe	British Leyland (U.K.)	Four robots spraying underseal.
	Fiat (Italy)	Four robots spraying final color at Cassino plant, plans to increase number to 20; two vertical and one horizontal reciprocating spray machines, four fixed guns, four robots, automatic door openers spraying plastic powder primer at Termini plant.
	Ford (Belgium/West Germany)	Two robots spraying underseal at Genk plant; one robot painting engines at Cologne plant.
	Renault (France)	One robot spraying corrosion-resistant primer on underbodies, reciprocating spray machines and three robots spray primer and final color, with manual "cut-in", at Flins plant. Total of 26 robots spraying paint, primer and undercoat.
	Rolls Royce (U.K.)	One robot spraying rustproofing on underbodies.
	Volkswagen (West Germany)	Automatic electrostatic primer and color coating, with manual interior spraying.
Asia	Honda (Japan)	Reciprocating spray machines and manual "cut-in" for primer and color coating. Sixteen robots for prime, color and underseal at Sayama plant.
	Mitsubishi (Japan)	Two robots spraying underseal and ten robots spraying color at Okazaki plant.
	Nissan (Japan)	Reciprocating spray machines and manual "cut-in" for primer and color at Zama plant; three robots spraying underseal at Oppama plant; 50 robots used for spray painting in various plants.
	Toyota (Japan)	Fixed-gun electrostatic, reciprocating spray machines and manual "cut-in" for primer and color.

-8-

Current Applications in Trim and Final Assembly

Trim and final assembly operations are still labor-intensive throughout the industry. With few exceptions, these operations are done on moving conveyor lines. Table 20 shows several approaches to automating these operations or to otherwise improving their efficiency.

TABLE 20 - AUTOMATION AND OTHER TECHNOLOGY UTILIZATION IN TRIM AND FINAL ASSEMBLY OPERATIONS

Region	Company (Country)	Technology
North America	Ford (U.S.)	Computerized electronic tester for instrument panel - checks 65 to 98 circuits in two minutes. For Escort/ Lynx carlines.
Western Europe	Fiat (Italy)	"Group" or "team" assembly of mechanical components, on fixtures in 12 pre-assembly bays, on self-propelled, wire-guided, computer-controlled carriers and automatic assembly of components to body in five automatic assembly machines, at Mirafiori plant.
	Volvo (Sweden)	"Group" or "team" assembly of trim components to body, chassis components on frame and final vehicle assembly on self-propelled, wire-guided, computer-controlled carriers, computer monitoring for quality and production control, at Kalmar plant.
Asia	Toyota (Japan)	"Stop" buttons installed at each station on assembly lines and workers are encouraged to stop the line any time they are unable to finish their task; has eliminated all on-line quality control inspectors. "Kanban" production control system which reduces in-process storage, floor space requirements, inventory and material handling requirements by frequent deliveries of small quantities of parts and material to the point of use.

foregoing sections of the report have shown, automobile manufacturing involves operations which range from making steel, glass, paints and plastics through casting, forging, stamping and machining to welding, assembly, painting and inspection. In all but the most basic operations, industrial robots are likely to be found.

The following table shows the approximate distribution of the automobile industry's robots among the significant applications and further, the distribution of robot types among these applications. The total population is for the year-end 1980 and the distribution is a best estimate; numbers for any specific application and/or type should be considered only as representative and should not be taken as absolute.

TABLE 21 - DISTRIBUTION OF ROBOT APPLICATIONS AMONG AUTOMOBILE MANUFACTURING OPERATIONS, YEAR-END 1980

	North America		Western Europe		Asia		TOTAL	
Application	Type	Number	Type	Number	Type	Number	Type	Number
Spot Welding	A	671	A	1,228	A	1,255	A	3,154
Parts/Material Handling	A	88	A	38	A	62	A	188
	C	13	C	49	C	974	C	1,036
		101		87		1,036		1,224
Machine Tending	A	64	A	16	A	205	A	285
	C	6	C	46	C	1,100	C	1,152
					D	20	D	20
		70		62		1,325		1,457
Stamping Press Loading/Unloading	A	35	A	4	A	14	A	53
	C	8	C	17	C	650	C	675
		43		21		664		728
Component Assembly	A	11	A	4	A	68	A	83
	C	6	C	11	C	280	C	297
		17		15		348		380
Foundry/Forge	A	6	A	12	A	16	A	34
	B	4	B	12	B	-	B	16
					C	120	C	120
		10		24		136		170

TABLE 21 (Continued)

Application	North America Type	North America Number	Western Europe Type	Western Europe Number	Asia Type	Asia Number	TOTAL Type	TOTAL Number
Arc Welding	B	30	B	127	B	97	B	254
Spraying	B	22	B	58	B	93	B	173
Deburring	B	2	B	22	B	11	B	35
Die Casting	A	38	A	6	A	-	A	44
	C	47	C	17	C	75	C	139
			D	13	D	70	D	83
		85		36		145		266
Sorting/Inspection	A	10	A	6	A	11	A	27
	C	4	C	2	C	96	C	102
		14		8		107		129
TOTAL	A	923	A	1,314	A	1,631	A	3,868
	B	58	B	219	B	201	B	478
	C	84	C	142	C	3,295	C	3,521
	D	0	D	13	D	90	D	103
		1,065		1,688		5,217		7,970

There are several significant differences to note between robot utilization in North America, Western Europe and Asia, as shown in the preceeding table. First, more than 90% of the robots used in automobile manufacturing in North America and Western Europe are programmable, servo-controlled (Types A and B) units, whereas only about 35% of the Asian automobile industry's robots are of these types. One implication of this is clear: assuming comparable labor savings can accure with the use of either Types A and B or Types C and D robots and assuming that Types C and D robots cost about one-third as much as Types A and B robots, the Asian (that is, Japanese) automobile industry's utilization of capital for reduction of manufacturing costs should be significantly more effective than in North America or Western Europe.

Another difference is in robot utilization for body spot welding. Utilization for these operations in North America is about 63% and in Western Europe, about 72%. In Asia, by contrast, only about 24% of the robots are spot welding. This is relatively close to the overall percentage of automated (robot and machine/multiwelder) spot welding in North America -- 52%, and in Western Europe -- 72%, but dramatically different from Asia, where about 85% of all spot welding is automated. Here the implication is somewhat different, automatic welding equipment, other than robots (that is, press welders, machine welders and multiwelders), generally costs about one-third more than robot welding lines performing comparable tasks. Thus, the greater cost savings in Asian body shops resulting from a higher degree of automated welding is offset somewhat by the higher capital investment in equipment to eliminate manual welding operations.

Significant differences in robot utilization between Asia and the rest of the world are also evident in basic manufacturing and fabrication operations. In parts handling, machine tending, press loading, foundry, forge, deburring and die casting operations, the Asian automobile industry's robot utilization is two to four times as great as in North America or Western Europe (more than 63% in Asia, less than 29% in North America and less than 15% in Western Europe). This pattern of application is in total agreement with the Asian automobile industry's tendency to use the simpler robots; many basic manufacturing and fabrication operations do not require servo-controlled (Type A and B) robots.

-9-

Projected Trends of Robots and Automation

There is little doubt that automobile manufacturers worldwide intend to continue to introduce automation into their manufacturing processes, particularly for labor-intensive operations and for quality assurance. For the near term, at least, most automation systems will use proven, state-of-the-art technology rather than introduce radical new ideas.

Robot utilization will continue to grow. Pallet transfer lines for machining and assembly will become more efficient through the use of robots for loading and unloading and in-process gaging will be incorporated into more of these systems.

New production facilities are planned by many automobile manufacturers and these will incorporate more robots and automation.

Major captal expenditures to modernize facilities, to increase production capacity and to develop and produce new products are planned by many European and Japanese automobile manufacturers, as well as by North American manufacturers.

Following past practices, it is most likely that the automobile manufacturers will use these programs to increase the level of automation of their operations, as well.

Table 22 shows some of these capital expenditure programs.

TABLE 22 - NORTH AMERICAN, WESTERN EUROPEAN AND ASIAN CAPITAL EXPENDITURE PROGRAMS

Region	Company (Country)	Technology
North American	Chrysler (U.S.)	Over the next five years (1981-1985), Chrysler plans to spend $6.5 billion to bring to market at least two front wheel drive models each year. It will spend more than $75 million to change over their St. Louis Plant. (10)
	American Motors (U.S.)	AMC plans a seven year (1981-1987), $1.6 billion program, including a new Jeep (1983 model) and a new family of front wheel drive cars design by Renault. (10)
	Ford (U.S.)	Ford plans to spend about $2.5 billion annually through 1985 for plant conversions and for 22 major product programs, including four or five new vehicle platforms, ten new and modified engines, nine new and modified transmissions and eight new front wheel drive carlines. (10)
	General Motors (U.S.)	General Motors plans to spend about $32 billion in North America from 1980 through 1984, including three new assembly plants, conversion of two more assembly plants, construction and expansion of several component plants, as well as redesign of most of its products. (10)
	Honda America (U.S.)	Construction of a $200 million car-manufacturing plant is underway, with completion scheduled in 1982. The plant will have a capacity of 10,000 units per month. (10)
	Nissan USA (U.S.)	Construction is underway on a $500 million plant to produce 150,000 Datsun pickup trucks per year, beginning in August 1983. (10)

TABLE 22 (Continued)

Region	Company (Country)	Technology
N.America (Cont'd)	Volkswagen America (U.S.)	Conversion of a former missile plant in Sterling Heights, Michigan, is underway at a cost of $300 million. The plant is to be operational late in 1982. By 1985, another $400 million to $500 million will go into tooling, upgrading and expanding facilities. (10)
Western Europe	Daimler Benz (West Germany)	About $5 billion will be spend in the next five years to double production capacity and develop new engines and car models. (1)
	Fiat (Italy)	About $5 billion will be spent in the next four years to launch a new car model and increase market share. (1)
	Ford-Europe	About $5.8 billion will be spent through 1985 to develop new car and truck models and retool its plants for them. (2) A new $70 million plastic part manufacturing plant will be opened in West Germany in 1981. (3) About $5.3 billion will be spent through 1984 to expand production facilities. (3)
	General Motors	About $2.4 billion will be spent to increase manufacturing capacity, including an assembly plant and several components plants in Spain, a transmission plant and an engine plant in Austria, a components plant in France and a components plant in Northern Ireland. (4) (5)
	Volkswagen (West Germany)	About $5 billion will be spent from 1979 to 1981 on new products and retooling for them. (2)
Asia	Isuzu (Japan)	Production capacity will be increased from 420,000 units a year to 600,000 units a year within next two years. (6) $121 Million will be spent on expansion of production facilities. (7)

TABLE 22 (Continued)

Region	Company (Country)	Technology
Asia (Cont'd)	Toyo Kogyo (Japan)	$167 Million will be spent to build a new assembly plant with capacity to make 20,000 cars a month, to replace less-efficient assembly operations at the main plant. Will be operational by the fall of 1982. (8)
	Toyota (Japan)	New assembly plant, with capacity to make 10,000 units a month will be operational in spring of 1981. (9) A second new plant, with similar capacity may be built to start up in 1982. (6)

Sources:
(1) The Engineer, September 20, 1979, p. 15
(2) Wall Street Journal, September 12, 1979
(3) Wall Street Journal, January 16, 1980
(4) Wall Street Journal, February 21, 1980
(5) The Engineer, June 14, 1979, p. 9
(6) Wall Street Journal, December 12, 1979
(7) Wall Street Journal, November 13, 1979
(8) Wall Street Journal, December 23, 1980
(9) The Engineer, April 3, 1980, pp. 32-5
(10) Wards' Auto World, June, 1981, pp. 36-40

International Resource Development, Inc., in their survey, Industrial Robots in the 1980's, published in November, 1979, projected that the United States automobile industry would be purchasing at a level of $21 million worth of robots (400 units) in 1984 and $28 million worth (600 units) in 1989. Indumar, Inc., a market research company in Cincinnati, Ohio, in a private study in June, 1980, projected that robot purchases by the U.S. automobile industry in the next five years would amount to 1,085 units, about 25% of total robot sales. Frost & Sullivan, in their survey, The United States Industrial Robot Market, published in November, 1978, projected 1985 robot sales to the U.S. automobile industry at more than $53.8 million. At the 10th International Symposium on Industrial Robots in March, 1980, Mr. Joseph Engelberger, President of Unimation, the world's largest robot manufacturer, predicted that the robot population in the automobile industry worldwide would grow by 600-700 units in 1980 and then annually by 35% in the 1980's to reach approximately 50,000 by 1990.

Outside of the United States, the situation is much the same. The Japan Industrial Robot Association's (JIRA) 1978 report, The Present Status and the Future Outlook of Industrial Robot Utilization in Japan, November, 1978 indicated 1977 robot production value of 18 billion Yen and projected production valued at 290 billion Yen in 1985, with 30% to 35% being installed in the automobile industry. This same report indicated a trend toward replacement of multiwelders by robots because of product diversification and a shorter life cycle of car models. At the 10th ISIR also, the executive direction of JIRA reported that the annual production of robots in Japan in 1978 was about 10,000 units and 25 billion Yen, 114% of the previous year's production, and that the automobile industry accounted for 35%, in value, of the 1978 robot installations.

A recent report published by Daiwa Securities America Inc., placed the Japanese robot production in 1979 at 2,763 units with a value of $102 million and estimated the 1980 production at 3,200 units valued at $180 million. The same report projected Japan's 1985 robot production at 31,900 units with a value of $2.15 billion and 57,450 units worth $4.45 billion in 1990. In 1979, the automobile industry purchased about 38% of the Japanese robots and in 1980, the percentage is estimated at 30%. On this basis, the robot population in Japanese auto plants at year-end 1985 could be around 31,250 units.

In the North American automobile industry (and, most likely in the foreign auto industry as well) there will be, in addition to steady growth in robot population, a shift in applications, as shown in the following table:

TABLE 23 - PROJECTED INDUSTRIAL ROBOT APPLICATIONS IN THE NORTH AMERICAN AUTOMOBILE INDUSTRY, 1980--1990

Application Area	Percent of Total Automotive Robot Population				
	1980	1983	1985	1988	1990
Spot Welding and Arc Welding	65	45	35	25	20
Painting	2	8	9	10	11
Assembly	1	15	24	33	36
Fabricating and Manufacturing	20	22	24	26	28
Parts Handling and Transfer	12	10	8	6	5

The most significant changes in application through the decade will be the dramatic increase in robots for assembly operations, an increase in robots for painting and a leveling off of the robots in welding operations, the latter resulting from the eventual installation of robots in most automobile body assembly facilities. Painting robots will also eventually replace all manual spraying operations and level off. Successful application of robots in assembly will facilitate automation of some of the most labor intensive operations in automobile manufacturing with a great potential for labor and cost reduction.

Table 24 summarizes some of the significant specific robot programs now being planned or underway. These are mostly applications of Types A and B robots, with Type A predominating.

TABLE 24 - MAJOR FUTURE ROBOT UTILIZATION PROGRAMS

Region	Company (Country)	Technology
North America	Chrysler (U.S./Canada)	Robogate body framing system and 64 Type A robots for St. Louis Assembly Plant, with startup in 1981. (1)
	General Motors (U.S./Canada)	14,000 or more robots to be installed by 1990 in body assembly, parts assembly, machining, stamping, casting, painting and forging operations. (2)
Western Europe	Ford	Ford of Europe will add between 300 and 450 robots by 1982 at plants in Belgium and United Kingdom for new 1983 carline. (3)
	BMW	BMW has 192 addtional robots on order for delivery by the end of 1982. (4)
Asia	Toyota (Japan)	200 robots being installed in new Tahara factory during next twelve months and a further 200 in other factories during the following twelve months. (5) Unimation's licensee, Kawasaki has received an order from Toyota for 720 robots. (5)

Sources: 1) Chrysler Manufacturing Engineering, June 25, 1981
2) Automotive Industries, March 1981, pp. 41-45
3) Ford Manufacturing Staff, June 26, 1981
4) The Industrial Robot, Volume 7, No. 4, December 1980, p. 259; American Metal Markets/Metalworking News, May 18, 1981, p.21
5) Daiwa Securities America Inc. Report #25, July 28, 1981, p.24

The major U.S. automobile producer, General Motors, has publicized its robotics plans for the 1980's. The plan projects a robot population in General Motors North American Manufacturing and Assembly Operations of 14,000 by 1990.

The projected applications distribution for these robots is shown in Table 25.

TABLE 25 - PROJECTED ROBOT APPLICATIONS IN GENERAL MOTORS NORTH AMERICAN OPERATIONS

	Number of Robots in Use			
	1983	1985	1988	1990
Welding (Arc and Spot)	1,000	1,700	2,500	2,700
Painting	300	650	1,200	1,500
Assembly	675	1,200	3,200	5,000
Machine Loading	200	1,200	2,600	4,000
Parts Transfer	125	250	500	800
TOTAL	2,300	5,000	10,000	14,000

Source: Bache Institutional Research Newsletter, June 19, 1981, p.3
General Motors Manufacturing Staff Productivity Conference, November, 1980
Industrial Robots International, July 1981, p.2

Some specific future trends and plans in component manufacturing and assembly, in body assembly and in body finishing, can be seen. These are described on the following pages.

-10-

Future Trends in Component Manufacturing and Assembly

Future plans for automation in automobile component manufacturing and assembly will involve robots, pallet transfer lines and special-purpose assembly machines. Examples of new technology and future plans are shown in Table 26.

TABLE 26 - FUTURE PLANS AND TRENDS IN AUTOMATION OF COMPONENT MANUFACTURING AND ASSEMBLY

Region	Company (Country)	Technology
North America	General Motors (U.S.)	Electro-optical gaging system for 100% inspection of crankshaft on production line.
		Five pallet transfer machining lines for front wheel drive transmission cases with 25 Type A robots (5 per line) for automatic loading and unloading.
		Two pallet transfer machining lines for blocks, two pallet transfer machining lines for heads, six to eight transfer machines for crankshafts and camshafts, 3-loop automated assembly system, 24-stand computerized hot test, diagnostic and repair system for new 1.8 liter four-cylinder engine.
		Two automated machining, welding and assembly lines for J-Car rear axle assemblies (cross beams, control arms, spindle plates and brackets, suspension mounting brackets, spring seats, bushing sleeves and braces). The second line will use four Type B robots to weld external reinforcements to the cross beams; robots may be added to first line later. Type A robots will also load and unload some sections of the lines.
		Robot based automated assembly system with several small Type A robots, small Type C robots and rotary index table to go into production operation early in 1982.

TABLE 26 (Continued)

Region	Company (Country)	Technology
Western Europe	British Leyland (U.K.)	Sixteen Type B robots for arc welding of Land-Rover chassis frames will be installed in 1981; one Type B robot for arc welding of light-gage sheet metal will also be installed in 1981.
	Daimler Benz (West Germany)	38 arc welding robots on order for component welding.
	Renault (France)	Type A robot with vision system will transfer crankshafts from a shipping pallet to a grinding machine; will be installed in 1981.
		Renault has begun production of small Type A robots for component assembly. About 30 have been installed in-house in 1980 and the company intends to produce about 140 in 1981; about 75% of production has been for its own use.
Asia	Aisin Seiki (Japan)	11-Station flexible machining line with automatic loaders and unloaders to produce 15 different brake cylinders being developed.
	Hitachi Seiki (Japan)	45-Station transfer line for automatic drilling, assembly, lubrication and leak test of cylinder heads.

-11-

Future Trends in Body Assembly

The use of Type A robots for welding and inspection in body assembly operations will continue to show significant growth, as indicated in Table 27.

TABLE 27 - FUTURE PLANS AND TRENDS IN AUTOMATION OF BODY ASSEMBLY

Region	Company (Country)	Technology
North America	American Motors (U.S.)	20 Robots will be installed in Kenosha plant for production of Renault-designed front wheel drive products, startup in spring of 1982.
	Chrysler (U.S./Canada)	Robogate system with 64 Type A robots for body assembly at St. Louis plant in 1981.
	General Motors (U.S.)	Two robots with laser gaging system for J-Car body inspection at Lordstown plant; similar systems will be installed in four other plants by 1982.
		36 Type A robots for body shell welding on J-Car bodies at Lordstown and South Gate. Robogates and robots for body assembly at new plants in Orion Township, Michigan and Wentzville, Missouri and renovated plants in Leeds, Missouri; Baltimore; Janesville, Wisconsin; Arlington, Texas; Lansing, Michigan; Framingham, Massachusetts; Doraville, Georgia; Fremont, California; Toronto, Canada; Flint, Michigan; Lakewood, Georgia, (several hundred robots are involved); 20 robots for body shell welding at Norwood, Ohio plant.
	Nissan USA (U.S.)	About 100 robots will be installed in new U.S. plant for frame, body and cargo box welding with startup in 1983.
	Volkswagen of America (U.S.)	VW robots will be installed in Sterling Heights plant for mid-1982 startup.

TABLE 27 (Continued)

Region	Company (Country)	Technology
Western Europe	BMW (West Germany)	A total of 192 robots have been ordered for production of new Type 3 and Type 5 car bodies, with installation to be complete by the end of 1982.
	Ford	Between 300 and 450 robots will be installed in Genk, Belgium and Dagenham, U.K. for production of new carline in late 1982.
	Volvo (Sweden)	Two body framing lines and one body shell welding line with 32 robots now being installed.
Asia	Ford (Australia)	Twelve robots being installed for underbody spot welding; startup in late 1981.
	Nissan (Japan)	Nissan plans to buy 350 robots, mostly for welding, in the next few years.
	Suzuki (Japan)	34 Robots in body shop welding underbodies and bodysides for four different bodies.
	Toyota (Japan)	Toyota has ordered 720 spot welding robots from Kawasaki Heavy Industries -- 220 for delivery by March, 1981, 200 by March, 1982 and 300 by March, 1983. Kawasaki is delivering about 25 units a month.
		Toyoda Machine Works, a subsidiary of Toyota is producing about 1,000 modular robots a year, for use within Toyota manufacturing operations.

-12-

Future Trends in Body Finishing

The use of Type B robots for body finishing will show significant growth, as shown in Table 28.

General Motors has developed its own Type B robot, which it calls the NC Painter, and may produce them for internal use.

TABLE 28 - FUTURE PLANS AND TRENDS FOR BODY FINISHING

Region	Company (Country)	Technology
North America	General Motors (U.S.)	Eleven NC Painters now in operation in Doraville, Georgia plant. About 300 painting robots will be installed by 1983, most of which could be NC Painters.
	Nissan USA (U.S.)	A total of 37 robots will be installed in a completely automated paint system in the new U.S. plant.
Asia	Nissan (Japan)	Nissan plans to install about 300 robots for spraying operations (prime, underseal, color) in the next few years.

-13-

Manufacturing Cost/Automation Relationships

Four major components of the cost of an automobile are product engineering, facilities and tooling, material and labor. With present market and regulatory pressures, changeover to new models and introduction of new products are frequent, and, as a result, product engineering, facilities and tooling costs are high. Size and weight reduction and corrosion protection programs have brought about changes in materials which often have increased material costs.

At the moment, the best opportunity for cost containment or reduction, therefore, lies in reducing labor costs. While some labor cost reduction can result from new product engineering, the primary means is through automation of manufacturing operations.

Other reasons for automation may exist, as will be cited in this section of the report. These include increasing productivity or production capacity, improving quality, reducing scrap or rework costs and coping with absenteeism and work stoppages, all of which have an impact upon manufacturing costs.

No hard data can be cited on the relationship between automobile manufacturing cost and the level of automation. However, based upon specific examples in Table 29 which follows, the conclusion can be drawn that higher levels of automation result in lower manufacturing costs. The examples relate to automation technology presented in the "Current Applications" sections of the report and are referenced to tables in those sections.

TABLE 29 - MANUFACTURING COST/AUTOMATION RELATIONSHIPS

Technology	Reference	Advantages
Robots unload die casting machines	Table 12 Doeler-Jarvis	Robots reduce labor by 60%, increase production by 15%, reduce scrap by 5%.
Robots load and unload forging presses	Table 12 Alfa-Romeo	Robots reduce labor by 50%, increase production 5% to 20%.
Fully automatic, flaskless molding machine for malleable iron	Table 12 Castings	Fully automatic Disamatic flaskless molding machine reduces manpower from 15 to 1.
Robot handles parts through upsetter	Table 12 Tube Investments	Robot reduces labor from 3 to 1.
Manipulators load and unload gear-making machines	Table 13 Nissan	1,500 manipulators loading and unloading gear-making machines reduced labor by 900 people (25%). Other automation in Yoshiwara transmission plant reduced work force by 300 people and increased output 10,000 units a month.
Gear rolling for finishing operations	Table 13 British Leyland, Borg Warner Ford-Europe	Gear rolling instead of shaving for finishing lowered tooling costs and increased production by 2.5 times.
Fully automated in-line stamping system	Table 15 Bendix	System significantly reduces direct labor cost, reduces in-process inventory and increases productivity by 33%.
Robots handle stampings through presses	Table 15 Honda	7-Press line at Sayama Plant operates with four men instead of 28.
Robots arc weld stampings	Table 15 Nissan	Robots reduce labor from 12 to 2.
High-speed line for automatic assembly of fuel gages	Table 16 Nippondenso	High-speed automatic assembly line for fuel gages reduced operators from 125 to 25 people.

TABLE 29 (Continued)

Technology	Reference	Advantages
Robots spot weld body panels	Table 18 Fiat	Robots eliminate one man per shift per robot.
	Table 18 Volvo	Six robots on indexing transfer line reduce labor by 5.5 people per shift.
	Table 27 Suzuki	34 Robots in Kosai plant reduce labor from 320 to 250 people per day.
Multiwelders spot weld body panels	Table 18 Fiat	2-Stage multiwelder for body shell eliminated 40 workers per day.
Robogates, carriers and robots spot weld body panels	Table 18 Fiat	System operates with 25 men per shift (8 on the floor and the rest backup maintenance) instead of 125 men per shift for conventional, non-automated system.
Robots check dimensions of body shells	Table 18 Ford-U.S.	Robots check five to six bodies per hour, instead of one to two bodies per 8 hours; with same manpower.
Multiwelders and robots weld body panels	Table 18 Saab	Three multiwelder systems and ten robots weld subassemblies and body shells, producing 80,000 bodies a year with 32 direct laborers a day.
	Table 18 British Leyland	Multiwelders for underbody assembly make more than 600 spot welds. Thirteen men required, instead of 80. Robots and multiwelders for body shell assembly make more than 460 spot welds. 38 Men required instead of 138.
	Table 18 Nissan	At Zama plant, a total of 160 direct workers on two shifts produce 800 bodies per day. They handle parts, manually weld and maintain equipment. There are no indirect workers on the floor of the body shop.

Automation may be installed for reasons other than the cost of labor. In Sweden, for example group assembly and automated transfer of work, as typified by Volvo's Kalmar plant was instituted to reduce absenteeism and labor turnover (almost 20% at the main Volvo plant) and to improve product quality. Volvo's engine plant at Skovde also uses assembly teams; there, absenteeism and turnover is 10%, half the national average in Sweden, engine rework has dropped from 1% to less than 0.1% and the average time of 3.75 hours to build an engine is down from previous years.

In Italy and the United Kingdom automation is applied in critical manufacturing operations to reduce the impact of unauthorized work stoppages.

In Japan, automation is considered essential to meeting increasing production requirements with a labor force whose size is relatively fixed. Also in Japan, the lifetime employment policy still followed by larger companies makes long-term investment in automation more economically attractive than labor-intensive approaches.

-14-

Investment Levels for Robots and Automation

Although automation and other new technology can significantly reduce labor cost, increase productivity and solve labor problems, investment levels are often high, as shown in Table 30.

TABLE 30 - INVESTMENT LEVELS FOR ROBOTS AND AUTOMATION

Technology	Reference	Investment Level
Programmable transfer lines for engine components	Table 14 General Motors	Programmable (convertible) transfer lines cost 8% to 10% more than conventional transfer lines, which may cost from $10 million to $15 million.
High speed automatic assembly lines for automotive instrument and gages	Table 16 Nippondenso	High-speed automatic assembly line for batch assembly of fuel gages cost about $1 million.
Computer-controlled carriers and "group" assembly for engine manufacturing	Table 16 Volvo	Skovde engine plant cost about $50 million in 1974, 10% more than conventional plant of comparable capacity; produces 275,000 engines a year.
Robogates, carriers and robots for body welding	Table 18 Chrysler, Fiat, Table 19 Chrysler, General Motors	Robogate systems cost 30% more than multiwelders, transfer lines and robots and occupy twice as much floor space. Fiat's Robogate system with 27 robots at the Rivalta plant cost more than $20 million in 1977.
Multiwelder for spot welding body panels	Table 18 Fiat	2-Stage multiwelder for body shell cost $3 million in 1973.
Multiwelders and robots for body panel welding	Table 18 Saab	Hood welding line cost nearly $1 million, front end welding line cost $6 million, body framing line cost $3 million, robots cost over $1 million, in 1977.

TABLE 30 (Continued)

Technology	Reference	Investment Level
Computer-controlled carriers and "group" assembly for trim and final operations	Table 20 Volvo	Kalmar assembly plant cost about $20 million in 1973, about 10% more than conventional plant of the same size; produces 60,000 cars a year on two shifts. Conventional plan might produce 180,000 cars with same space.
Robots for body assembly, parts assembly, machining, stamping, casting and forging operations	Table 25 General Motors	General Motors' future robot purchases would cost between $70 million and $126 million, at today's prices, for 1,000 to 1,800 robots.

The actual population of robots in the North American automobile industry, by the end of the decade, could reach 35,000 units or more. Projections of this population growth are shown in the following table:

TABLE 31 - PROJECTED INDUSTRIAL ROBOT POPULATION IN THE NORTH AMERICAN AUTOMOBILE INDUSTRY, 1980-1990

	1980	1983	1985	1988	1990
Minimum Effort	1,065	2,600	4,700	10,800	18,500
Moderate Effort	1,065	4,050	7,500	16,200	22,600
Strong Effort	1,065	4,500	10,000	20,000	28,000
Maximum Effort	1,065	4,500	11,200	25,000	35,700

The estimated total cost of these robots, including engineering of the applications; systems development, design and fabrication; and implementation, but not including operating support expenses (that is, the initial investment only), in constant (1980) dollars, is shown in the following table:

TABLE 32 - CUMULATIVE COST OF INDUSTRIAL ROBOTS IN THE NORTH AMERICAN AUTOMOBILE INDUSTRY, 1980-1990 (Millions of Dollars)

	1980	1983	1985	1988	1990
Minimum Effort	120	300	550	1,200	2,000
Moderate Effort	120	450	850	1,800	2,500
Strong Effort	120	500	1,100	2,200	3,000
Maximum Effort	120	500	1,250	2,750	4,000

The "minimum effort" projections are based on the assumption that the domestic automobile industry will continue to experience lagging sales and strong foreign competition through the mid-1980's. Under such circumstances, the rate at which robots are added to the current population will not be accelerated from the present level and few new, innovative applications will be developed. The "moderate effort" projections assume some recovery of the domestic automobile market and a modest decline in interest rates, making more capital available for productivity improvements. The "strong effort" projections are based upon a similar rate of recovery of the market and interest rate decline, plus investment incentives such as tax credits, accelerated write-offs, etc., and technological advances such as low-cost sensory feedback systems. The "maximum effort" projections assume, in addition to the foregoing, direct subsidies for robot investments, government-funded worker retraining and/or relocation programs and other direct incentives.

Adv

-15-

Potential Benefits from Robots

The primary benefit of industrial robot implementation is a reduction in the labor cost to produce a finished product. In some automobile manufacturing operations, added benefits such as energy and material savings (spray painting), more efficient utilization of assets (machine tending), reduction of scrap (die casting, plastic molding, etc.) and warranty cost reduction (in-process inspection) may accrue. These are, however, of significantly lesser magnitude than the labor cost savings.

Estimates have been made of the potential savings which the North American automobile industry can expect from the continued introduction of robots at various rates (as in Table 31). These estimates, which are shown in the following table, are based upon a number of factors, including labor savings, energy and material savings, higher output and reduced scrap and rework. Improvements in robot performance and efficiency, changes in applications patterns and hours of utilization were also considered.

In order to maintain production while individual workers on the line are taking their authorized breaks ("relief"), normal practice in North American auto factories is to temporarily replace these workers with "utility" or "relief" operators. Thus, for approximately every seven workers in production operations, there is an additional worker required. Robots, obviously do not require relief, therefore, each robot replaces not only the direct labor operators, but an increment of a relief operator. Robots, however, do require maintenance and repair. Thus, for each robot installed, there is an incremental increase in the indirect labor force in the plant. For these estimates, the relief allowance savings and maintenance allowance increase were considered to offset each other.

Not considered in the savings, however, were the normal "overmanning" allowance, which is included in work standards to cover authorized and unauthorized absenteeism or an allowance to cover the lower efficiency of temporary or permanent replacement workers.

Wage and fringe benefit costs used were based upon average 1980-1981 rates in the U.S. auto industry, in constant (1980) dollars (as were the estimated robot costs in the previous section).

TABLE 33 - AVERAGE ANNUAL SAVINGS PER ROBOT IN NORTH AMERICAN AUTOMOBILE INDUSTRY, 1980-1990 (Thousands of Dollars)

	1980	1983	1985	1988	1990
Direct Labor Savings per Robot/per Shift a/					
- Heads	0.9	0.9	1.0	1.1	1.2
- Dollars	34,700	34,700	38,600	42,400	46,300
Robot Utilization b/					
- 2-Shift (Percent)	70	55	45	35	30
- 3-Shift (Percent)	30	45	55	65	70
"Other Savings" Factor (Dollars) c/	7,000	8,000	10,000	12,000	13,000
Average Annual Savings per Robot	86.8	93.0	108.4	124.3	138.0

a/ Assumes increasing efficiency of robots through technological improvements, sensory feedback, etc.

b/ Spot welding, arc welding and painting are associated with car assembly operations which operate on 2 shifts a day; other applications are associated with basic manufacturing operations which operate on 3 shifts a day. The change in shift utilization reflects the change in applications as shown in Table 23.

c/ Energy and material (painting); higher output, reduced scrap and rework (fabrication and manufacturing) -- factor varies with percentage of utilization of robots for these operations.

Based upon the projected population of robots, under various conditions, as shown in Table 31, and the average savings per robot in Table 33, the estimated cumulative savings are shown in the following Table:

TABLE 34 - CUMULATIVE SAVINGS FROM INDUSTRIAL ROBOTS IN NORTH AMERICAN AUTOMOBILE INDUSTRY, 1980-1990 (Millions of Dollars)

	1980*	1983	1985	1988	1990
Minimum Effort	390	890	1,725	4,600	8,900
Moderate Effort	390	1,100	2,450	6,900	12,500
Strong Effort	390	1,150	2,900	8,500	15,000
Maximum Effort	390	1,150	3,100	10,000	18,500

*Including estimated cumulative savings prior to 1980

Another, perhaps more meaningful, expression of the estimated savings from robot implementation is contained in the following table which shows the approximate savings per car. These estimates are based upon the same total savings as in the previous table and assume an annual production rate ranging from 7.2 million units in 1980 to 8.6 million units in 1990. The estimates are in constant (1980) dollars.

TABLE 35 - SAVINGS PER CAR FROM INDUSTRIAL ROBOTS IN NORTH AMERICAN AUTOMOBILE INDUSTRY, 1980-1990 (Dollars per Unit)

	1980	1983	1985	1988	1990
Minimum Effort	12	33	67	158	297
Moderate Effort	12	52	107	237	363
Strong Effort	12	57	143	292	449
Maximum Effort	12	57	160	366	573

Industrial robots also provide other, less tangible, benefits to the automobile industry. Robot spot welding systems are an excellent example. Body sheet metal welding is one of the most costly operations in automobile manufacturing. It requires a large investment in facilities, including electrical power, compressed air and cooling water distribution; welding transformers and controls; special-purpose automation; conveyors and transporters. It also requires a large investment in tooling, including assembly jigs and fixtures, welding machines, welding fixtures and welding guns. Much of this tooling has a short life; it is either replaced or extensively revised whenever styling or model changes are made. At the present time, although annual styling changes are no longer common, "downsizing", new model introductions and shifts in production rates are causing frequent, major changes in body tooling and facilities in many plants.

There are basically three approaches to body assembly: minimum investment, automation and robotics. The minimum investment approach uses simple, low cost jigs and fixtures and mostly manual welding. This method is inefficient because of high direct labor costs and excessive handling of components from fixture to fixture (and/or extensive conveyor systems for handling). Quality tends to be poor because of the high reliance upon operators to properly position, clamp and weld parts; because of the multitude of fixtures which require continual repair and adjustment to maintain the proper interrelationship of parts; and because of frequent handling of components.

The automation approach requires a large capital investment and has essentially no salvage value. The method is efficient in that a minimum of direct labor is required. Quality is potentially high, but may be adversely affected by the complexity of the automation which is often difficult to adjust and maintain. In today's time of frequent, major changes in body tooling, the investment in the automation may not be recovered before it must be either scrapped or extensively reworked. Installation of new automation or changeover of automation on-site is generally a time-consuming task, requiring long shut-downs of plants to accomplish.

The robotic approach may not be as efficient as automation in terms of direct labor requirements or in utilization of floor space, but its inherent flexibility and reliability make it an attractive alternative. Quality is potentially high and robotic systems are generally easier to adjust and maintain than automation. Robots do not weld as quickly as automatic welders or, in fact, as quickly as manual welders, thus, the initial investment may approach that of automatic welders. However, the robots are not obsoleted by model or styling changes and modern robot welding systems are designed so that most of the associated jigs, fixtures and transport mechanisms are also salvageable. Changeover time for new models is also relatively brief.

The concept of programmable automation, as typified by industrial robots, is also being adopted for other manufacturing processes. Several "convertible" transfer machining lines have been recently installed in two domestic engine plants. These lines are designed to produce several related products with only simple adjustments and replacement of a few heads or cutting tools involved for changeover. Normally a conversion can involve up to 90% of the cost of a new line and several months downtime. These programmable transfer lines may cost 10% more than conventional lines, but the overall savings are significant.

-16-

Impacts of Robots and Automation

The continuing introduction of robots will have an impact upon several classes of workers in the automobile industry, including unskilled laborers (that is, direct labor), skilled trades personnel (that is, maintenance personnel) and engineering personnel. Jobs will be lost in one sector and gained in others. A surplus in certain labor classifications will exist, concurrent with shortages in other labor classifications.

The most severe impact will be upon the direct labor force. Each robot installed today displaces about two workers a day. As robot's capabilities increase and applications shift to the manufacturing areas which tend toward 3-shift operations, the ratio of robots to workers displaced will increase, so that by 1990 each robot will displace more than three workers a day.

Offsetting the loss of direct labor positions somewhat is the requirement for maintenance and repair of the robots. On average, a robot requires about one-tenth of a manhour of maintenance and repair for each hour of operation. Thus, every ten robots installed in a factory require approximately one skilled trades person for each shift of operation. Exact ratios vary with the types of robots, severity of operations, numbers of robots working together or in related applications, etc. Although the robot's operating capabilities are expected to increase during this decade, implying more complexity and higher maintenance requirements, their reliability and self-diagnostic abilities will increase also. Thus, this ratio (unlike the direct labor ratio) is expected to remain relatively constant.

The loss of direct labor opportunities, the gain of skilled trades positions and the net effect upon hourly work availability is shown in the following table, for various rates of robot introduction. The robot population figures, which are included in the table for reference, and the levels of effort are as described in Table 31 and its supporting detail.

automation is a continuing problem. We will struggle with the human results of it". He added, "Automation is the way we have improved the standard of living of the workforce".

The UAW has been pressing the auto industry to provide retraining for workers affected by robots. It will also press for more job protections. Ephlin pointed out that Japanese auto workers, "have achieved somethings we haven't" - notably guaranteed job security and wages that are rising faster than the rate of inflation. He pointed out that the Japanese economy is growing so rapidly it allows the country to achieve its national goal of full employment while increasing its level of automation. "I don't see any national commitment to full employment here", he said. "1948 was the last time we looked at full employment. It's time we dusted it off."

Japanese auto workers and the Japanese automobile industry do not face the same set of circumstances as their North American counterparts. The Japanese employes in major corporations are guaranteed lifetime employment. In addition, all employes received two bonuses a year, which are based upon the company profitability. Japanese unions are not based on crafts, skills or occupations; a union is on a companywide basis and covers all members of the bargaining unit. Employes identify with the company, not with a skill and they are often shifted from one job to another within the company. The worker, not fearing loss of employment does not oppose automation; in addition, as automated production generally enhances quality and profit and consequently the bonus, the Japanese employes welcome the robots. Also, because robots are often used in dangerous, unhealthy and repetitious jobs, the employes consider production by robots as a good means of relieving monotonous and environmentally harmful tasks in manufacturing. Finally, the shortage of labor and the aging of the workforce will hasten the acceptance and implementation of industrial robots.

In Western Europe, the automobile industry is currently in a similar condition to North America's. The European labor unions' attitudes toward robotics vary widely from country to country, ranging from strongly supportive in Sweden through lesser degrees of support and opposition in West Germany, France, Italy and Belgium to rather strong opposition in the United Kingdom. However, despite lower sales and higher unemployment, the European automobile industry is implementing robot applications at a rate comparable to the rest of the world.

The increase in requirements for skilled trades personnel for robot repair and maintenance is creating a need for facilities and programs to provide basic training in several fundamental areas. Among these are fluid mechanics (pneumatics and hydraulics), mechanisms and electronics. Although specialized training in trouble-shooting and repair of specific robots is provided by either the manufacturer or the user of the robots, fundamental skills are also required. Currently, the North American automobile industry is experiencing a shortage of trainable skilled trades personnel to support their robot operations. The problem is compounded by the job/skills classification system in the industry which requires as many as three different maintenance personnel to remove and replace a defective component on a robot.

In the long term, the development of company-sponsored and/or company-administered basic or fundamental training programs may be required. The establishment of a multi-skill "robot repairman" job classification will also be necessary for cost effective robot maintenance. It is highly unlikely, however, that direct labor workers in North America will ever be trained and allowed to operate and maintain the robots associated with their tasks, as is the case in Japan.

There will be several impacts on the robot industry, as the North American automobile industry continues its implementation of robot applications. The most immediate of these impacts is on the ability of the robot manufacturers and distributors to deliver the desired quantities of robots in the time frames required, without diverting units from other customers.

Currently, the North American automobile industry purchases about one-fourth of all the robots manufactured or distributed here. If one assumes that this proportion of total output continues to be supplied to the auto industry, there could be shortages, based upon robot industry projections of production and sales. The following table shows projected robot industry output and automobile industry demand:

TABLE 37 - PROJECTED INDUSTRIAL ROBOT PRODUCTION AND SALES TO NORTH AMERICAN AUTOMOBILE INDUSTRY, 1980--1990

		1980	1983	1985	1988	1990
Total Annual U.S. Robot Production/Sales	a/	1,269	3,122	5,690	14,000	25,515
Annual Sales to North American Auto Industry	b/	317	780	1,421	3,497	6,379
New Robots Required - Minimum Effort	c/	0	900	1,200	2,600	4,400
New Robots Required - Moderate Effort	c/	0	1,450	1,800	3,300	3,200
New Robots Required - Strong Effort	c/	0	1,450	2,900	3,900	5,000
New Robots Required - Maximum Effort	c/	0	1,900	3,400	5,000	5,300

a/ Robot Institute of America forecast
b/ Assuming about 25% of total production
c/ From projections in Table 31

As the table indicates, the robot industry's apparent capacity to supply the automotive market falls short, particularly in the mid-1980's, under even the least demanding of implementation schedules. It is unlikely, however, that actual shortages of robots will exist. Additional manufacturers and distributors of robots are likely to emerge in the North American market and existing manufacturers and distributors may also temporarily allocate a greater portion of their output to the automotive industry.

The most likely new source of robots in North America will be Japan. At the moment, there are nine European and six Japanese robots, out of a total of twenty-eight brands, being marketed and/or manufactured in the U.S. The Japanese robot industry is already planning on significant exports. The annual production volume of their industry is projected to increase ten-fold between 1980 and 1985 to about 32,000 units and to almost double again between 1985 and 1990 to more than 57,000 units. About 16%, by value, of 1985 production is slated for export and about 20%, by value, of 1990 production (JIRA projections).

Thus, a secondary impact on the North American robot industry will be on its competitive position in one of its major home markets. In order to retain its sales in the important automotive market, the domestic robot industry will have to match the price, delivery, reliability, support and capability of the imports and keep pace with emerging technology as well. This latter requirement may be most difficult, in light of the comparative levels of effort in robotics research and development here and overseas, as described in an earlier section of this report.

Another impact will be on the robot industry's ability to provide support services to its customers. As previously mentioned, the automobile industry has reduced its staff of robot applications personnel. The automobile industry's requirements for skilled trades personnel and technicians is, at the same time, increasing. On the one hand, this means that the robot industry will be expected to provide applications and system engineering services, a role which that industry was not previously required to perform for the auto makers. On the other hand, the robot industry will have to compete with the automobile industry for technicians and skilled tradesmen for their manufacturing and field service/support operations. Presently, due to a wide disparity in pay and benefits, the robot industry is losing qualified personnel to the automobile industry. Matching the wage and benefit levels of the automobile industry and maintaining an applications and system engineering support staff could prove very costly to the robot industry at a time when competition from overseas requires holding the line on prices and increasing expenditures for research and development.

Internally, foreign automobile manufacturers are facing many of the same problems as the domestic industry. As with the domestic automobile industry, however, these problems are not considered as serious deterrents to continued implementation of robots. No significant impact is seen on those foreign competitors using robots as a result of the North American automobile industry's robotics utilization, since comparable benefits will accrue to both. Naturally, those not using robots, or using robots to a significantly lesser degree are operating under a cost penalty.

As the North American automobile industry increases its utilization of robots and automation, the productivity of the industry should continue to increase. At the moment, the annual rate of increase in productivity averages about 3%. By comparison, the Japanese automobile industry has an average annual productivity increase of 6%. Because the Japanese are introducing robots and automation into their automobile manufacturing operations at a rate at least comparable to the United States, there is little likelihood that the U.S. will match or exceed Japan's productivity. A similar comparison holds true for much of the European automobile industry, as well. The net result is that the U.S. automobile manufacturers will have to continue to implement robotics and automation at a strong rate in order to just keep up with foreign competition.

As with foreign competition, the impact of robotics on domestic automobile manufacturers will relate to their respective levels of utilization. Currently, no one domestic automaker (except AMC) is at a significantly higher or lower level of robot use than any other. Announced plans, internal planning and industry observer projections all point to the domestic automakers and the foreign automakers with domestic manufacturing operations continuing to implement robots at rates which will keep them competitive.

In addition to providing a means to improve productivity, industrial robots have an impact upon the type of tooling installed in the auto plants (especially in the assembly plants) the planning of changeovers and, to some extent, the layout of the plants themselves. As described earlier, the flexibility of an industrial robot is of advantage for model changeover. An analysis and comparison of hard automation and robot welding systems on the same operation at two different plants has been made by a domestic automaker. This comparison indicated that the initial cost of the robot system was about two-thirds of the cost of the hard automation system and that the cost for a subsequent model changeover was about half. In addition, the lead time for the robot system retooling for the changeover was about two-thirds that of the hard automation and the actual changeover time was about one-third less.

The robot's flexibility is also advantageous for coping with sourcing changes, that is, the major changeover of a plant to produce a completely different carline. The reusability of the robot is particularly important here. For example, when Chrysler stopped production at the Lynch Road Plant, the fourteen robots which were welding there were put into storage. These robots are now being shipped to Chrysler's St. Louis Plant where they will join some robots already in use there and some new robots being purchased as part of a new welding system which goes into operation late in 1981. The changeover of the robots from Lynch Road will involve simply installing new welding guns on their arms and putting new instructions (programs) into their memories.

As more and more "flexible automation" (that is, robots) is installed, the cost of retooling for model or sourcing changes should become less. Automotive industry manufacturing personnel experienced with robotic systems estimate that as much as 20% of the normal cost of retooling an assembly plant body shop can be saved where robots are being used. In addition to savings in tooling, the typical changeover time can be reduced by about one-third. Assuming that a major model change in a single assembly plant body shop equipped with hard automotion can cost as much as $40 million and require several weeks of downtime, the robots could save more than $10 million.

The robotics approach will also affect the design and layout of assembly plants. For maximum flexibility and efficiency, guided-transport systems and robotic work cells may replace continuous moving convcyor lines, not only in the body shop, but in trim and final assembly areas as well. This approach may require as much as twice the floor space as a conventional plant of comparable production capacity. This was Fiat's experience with their "Robogate" installation in the Rivalta plant and Volvo's at the Kalmar assembly plant, which uses guided-transport systems throughout and requires nearly three times the floor space of a conventional flow-line (moving conveyor) plant of comparable production capacity. Industrial plant and material handling systems designers also estimate that floor space requirements and facility costs for component manufacturing plants may be 10% higher for the robots and guided-transport systems approach.

As previously reported, estimates of the cost to the North American automobile industry to purchase and install robots range from about a half billion to a billion and a quarter dollars by 1985 and from two to four billion by 1990. These expenditures are in addition to more than $42 billion which the domestic automobile industry plans to spend through 1985 for new plants and new products.

Although these robot program expenditures are modest in comparison to the overall spending level, they are still significant. They also tend to be discretionary programs, that is, the delay or cancellation of a robot program usually has little or no immediate negative impact and may, in fact, make capital available for other programs. The long-term effect of such a delay or cancellation, however, as was discussed previously, will be the loss of competitive position or advantage.

-17-

Conclusions

The automobile industry worldwide utilizes, in general, the same types and technology levels of automation in manufacturing operations. No significant technological developments appear to be unique to a particular manufacturer or region. Obvious differences in the use of specific automation technologies and in the overall level of automation utilization do, however, exist. These differences are related to a number of factors, including labor rates and availability of workers, product complexity and differentiation and levels of investment and investment priorities. Direct labor is a significant element in the cost of manufacturing of automobiles and automotive components. The introduction of automation can greatly reduce labor costs, thus, higher levels of automation should result in lower cost of manufacturing the product.

In general, the degree of automation of automobile component manufacturing appears to be about the same, worldwide. Body assembly (welding) operations are more automated in Western Europe and Asia (Japan) than in North America, as are body finishing operations. Final assembly operations are similar worldwide, with a few unique exceptions.

Automation and robotics levels will continue to rise throughout the automobile industry worldwide through the 1980's. The growth rate of the robot population will be in the range of 35 to 40 percent a year through the decade. By 1990, the robot population in the North American automobile industry could reach 35,000 units or more, at a cost of $4 billion. The total savings in manufacturing costs, at that level of robot utilization, would be around $18 billion. This continuing implementation of robots will have an impact upon the labor force. By 1990, more than 115,000 direct labor positions in the North American automobile industry could be taken over by robots; about 10,000 new skilled trades/technician positions would be generated.

Robots and automation are effective tools for improving productivity and quality. Thus, despite the cost and the impact upon employment opportunities, their use will continue to expand. At this time, there are no technological factors inhibiting robot implementation. The development and refinement of new technology to enhance robots' capabilities remain comfortably ahead of the need for such technology. For the short term (the next two to four years), the inhibiting factors will be the users' lack of experience and confidence, a shortage of capital and, in North America, a limited robot manufacturing capacity. Beyond the mid-1980's, however, most of these hinderances will disappear.

References

American Machinist, McGraw-Hill, Inc. Hightstown, New Jersey

American Metal Market/Metalworking News, Manasquan, New Jersey

Assembly Engineering

Automotive Engineering, Society of Automotive Engineers, Warrendale, Pennsylvania

Automotive Industries, Chilton Co., Radner, Pennsylvania

Detroit Engineer, Engineering Society of Detroit, Detroit, Michigan

The Engineer, Morgan-Grampian Ltd., London, England

Engineering, Design Council, London, England

Ford Motor Company - Japan Trip Report, 1975

Ford Motor Company - "Universal Transfer Devices" (internal report), 1980

General Motors - "Industrial Robot Applications within General Motors" (internal report), 1976

IEEE Spectrum, Institute of Electrical and Electronics Engineers, Piscataway, New Jersey

Industrial Finishing, Hitchcock Publishing Company, Wheaton, Illinois

The Industrial Robot, IFS (Publications) Ltd., Bedford, England

Industrial Robots International, Technical Insights, Inc., Fort Lee, New Jersey

Iron Age, Chilton Co., Radnor, Pennsylvania

Institutional Research, Bache Halsey Stuart Shields Inc., New York, New York

Machine Design, Penton Publishing Company, Cleveland, Ohio

Manufacturing Engineering, Society of Manufacturing Engineerings, Dearborn, Michigan

Mechanical Engineering, American Society of Mechanical Engineers, New York, New York

Modern Machine Shop, Gardner Publications Inc., Cincinnati, Ohio

Modern Material Handling, Cahners Publishing Company, Boston, Massachusetts

Paul Aron Report #25, Daiwa Securities America Inc., New York, New York

"The Present Status and the Future Outlook of Industrial Robot Utilization in Japan", Japan Industrial Robot Association, Tokyo, Japan

Proceedings of the 9th International Symposium on Industrial Robots, Society of Manufacturing Engineers, Dearborn, Michigan-1979

Proceedings of the 10th International Symposium on Industrial Robots, IFS (Publications) Ltd., Bedford, England

REFERENCES (Continued)

Production, Huebner Publications, Inc., Solon, Ohio

Production Engineering, Penton Publishing Company, Cleveland, Ohio

Regie Renault - "Usine Pierre Lefaucheux" (internal report), 1980

Robot Institute of America - "Preliminary Results of Worldwide Survey", SME, Dearborn, Michigan, 1980

Robot Institute of America - "National Coordinator's Presentation", SME, Dearborn, Michigan, 1979

Robotics Today, Robotics International of SME, Dearborn, Michigan

Robots in Industry, Unimation, Inc., Danbury, Connecticut

Today's Machine Tool Industry, News Digest Publishing Company, Tokyo, Japan

The Wall Street Journal, Dow Jones & Company, New York, New York

Ward's Auto World, Ward's Communications Inc., Detroit, Michigan